Engineering Science

Engineering Science
Questions and Answers

L. J. Beeching, CEng, FIMarE, MRINA

Foreword by
Don Ewart, CEng, MIMarE, MRINA
Editor–Fairplay International Shipping Journal

STANFORD MARITIME LONDON

Stanford Maritime Limited
Member Company of the George Philip Group
12-14 Long Acre London WC2E 9LP

First published 1975
© 1975 L. J. Beeching

Printed in Great Britain by
J. W. Arrowsmith Limited Bristol

ISBN 0 540 07341 5

Foreword

Written by experienced lecturers at one of Britain's leading marine engineering colleges each book of this series is concerned with a subject in the syllabus for the examination for the Second Class Certificate of Competency. It is intended that the books should supplement the standard text books by providing engineers with numerous worked examples as well as easily understood descriptions of equipment and methods of operation. Extensive use is made of the question and answer technique and specially selected illustrations enable the reader to understand and remember important machinery details.

While the books form an important basis for pre-examination study they may also be used for revision purposes by engineers studying for the First Class Certificate of Competency.

Long experience in the operation of correspondence courses has ensured that the authors treat their subjects in a concise and simple manner suitable for individual study—an important feature for engineers studying at sea.

Don Ewart

Preface

The aim of this book is to give the student of Engineering Science suitable questions to test his knowledge of the subject. Although no claim is made that these are examination questions, they are designed to reflect the examination standard. Many of the questions have been suggested by students returning from examination and the author is indebted to those students for their contributions towards this volume.

Grateful thanks are also due to Mr John Palmer for correcting this work.

A list of definitions and formulae has been included to remind the student of the basic form of the fundamental equations in this subject. The units in which these equations should be worked are given in brackets; the dash (—) indicating a ratio or coefficient having no units.

An alphabetical list of the symbols used with their preferred meanings is given together with tables of units and factors at the front of the book.

No attempt has been made to explain any of the mathematical methods employed in the worked examples. For a more detailed description of these methods reference should be made to *Mathematics* by P. T. Yeandle in this series.

L. J. Beeching
Sidcup, Kent

Contents

SI Units

Basic quantity	Basic unit	Abbreviation
Mass	kilogramme	kg
Time	second	s
Length	metre	m
Angle	radian	rad

DERIVED UNITS

Quantity	Unit	Abbreviation
Force	Newton	$N = \dfrac{kgm}{s^2}$
Work	Joule	$J = Nm$
Energy	Joule	$J = Nm$
Power	Watt	$W = \dfrac{Nm}{s}$
Pressure	Newton/square metre	N/m^2
Stress	Newton/square metre	N/m^2
Density	kilogramme/cubic metre	kg/m^3
Area	square metre	m^2
Volume	cubic metre	m^3
Velocity	metre/second	m/s
Acceleration	metre/second/second	m/s^2
Angular velocity	radian/second	rad/s
Angular acceleration	radian/second/second	rad/s^2

PREFERRED FACTORS

Value	Prefix	Abbreviation
10^9	giga	G
10^6	mega	M
10^3	kilo	k
10^{-3}	milli	m

NOTES

The preferred factors may be applied directly to many of the units used in Engineering Science. However, there are a number of exceptions.

Mass. The kilogramme is 10^3 grammes (g) thus care must be exercised when applying factors to units of mass.

$$1Mg = 10^6g = 10^3kg$$
$$1g = 10^0g = 10^{-3}kg$$
$$1mg = 10^{-3}g = 10^{-6}kg$$

Time. The submultiplier *milli* is commonly used, however decimal multipliers are not available. Use is made of the minute (min), hour (h), day (d), etc.

$$1min = 60s$$
$$1h = 3600s$$

Angles. Large angles of displacement are measured in revolutions (rev) and it is common practice to measure small angles in degrees (°), minutes ('), seconds (").

$$1rev = 360° = 2\pi rad$$
$$1° = 60' = 3600''$$

Summary of the Symbols used in Engineering Science

The following symbols are given in alphabetical order (small case first) with the basic units in brackets and then the generally accepted meanings. When more than one meaning applies each is given.

a	(m/s^2)	linear acceleration
A	(m^2)	area
b	(m)	breadth—linear distance
B	(m)	breadth—linear distance
d	$(—)$	relative density
d	(m)	diameter—linear distance
D	(m)	diameter—linear distance
e		earth point
E	(N/m^2)	modulus of elasticity
E	(N)	equilibrant force—effort
F	(N)	accelerating force
F	(N)	frictional resistance—force
g	(m/s^2)	gravitational acceleration, taken as $9.81 m/s^2$
G	(N/m^2)	modulus of rigidity
h	(m)	head—linear distance
H	(m)	depth of centroid—linear distance
I	(m^4)	2nd moment of area
J	(m^4)	polar 2nd moment of area
k	(m)	radius of gyration
l	(m)	length—linear distance
L	(N)	load—force
m	(kg)	mass
M	(Nm)	bending moment (BM)
n	(rev/s)	speed of rotation
N	(N)	normal reaction—force
p	(N/m^2)	pressure
P	(W)	power
r	(m)	radius—linear distance
R	(m)	radius of curvature
R	(N)	reaction—force
R	(N)	resultant force
s	(m)	linear distance travelled
t	(s)	time
T	(Nm)	torque, turning moment
v	(m/s)	linear velocity
V	(m^3)	volume
W	(J)	work

3

W		(N)	weight—gravitational force
x		(m)	linear distance
X		(m)	distance to centroid
y		(m)	linear distance
Y		(m)	distance to centroid
z		(m)	linear distance
(alpha)	α	(/°C)	coefficient of linear expansion
(alpha)	α	(rad/s)	angular acceleration
(alpha)	α	(rad or °)	angle
(beta)	β	(°)	angle
(gamma)	γ	(—)	shear strain
(delta)	δ, Δ	(—)	change of
(epsilon)	ε	(—)	direct strain
(eta)	η	(—)	efficiency
(theta)	θ	(rad or °)	angle—angular displacement
(theta)	θ	(°C)	temperature
(mu)	μ	(—)	coefficient of friction
(pi)	π	(—)	a constant = 3·142
(rho)	ρ	(kg/m³)	density
(sigma)	σ	(N/m²)	direct stress—bending stress
(sigma)	Σ	(—)	sum of
(tau)	τ	(N/m²)	shear stress
(phi)	ϕ	(rad or °)	angle—friction angle
(omega)	ω	(rad/s)	angular velocity

This list does not impose any restriction on the use of other symbols which may be appropriate to the solution of a particular problem.

Definitions and Equations

Fundamental Force Equation

$F = ma$
- F = applied force   (N)
- m = mass   (kg)
- a = acceleration (m/s^2)

Weight—Gravitational Force

$W = mg$
- W = gravitational force   (N)
- m = mass   (kg)
- g = gravitational acceleration (m/s^2)

Moment of Force $= Fx$

- F = applied force   (N)
- x = perpendicular distance   (m)

STATICS

Equilibrium is the state of rest or uniform motion in a straight line.

Condition for Equilibrium

Nett sum of forces in one plane	= zero
Nett sum of forces in perpendicular plane	= zero
Nett sum of moments of force	= zero

Sign Convention for Equilibrium Conditions

Upward forces, forces to right, anticlockwise moments—POSITIVE

Downward forces, forces to left, clockwise moments—NEGATIVE

Equations Derived from Conditions for Equilibrium

Forces up	= Forces down
Forces to right	= Forces to left
Anticlockwise moments	= Clockwise moments

The planes above may rotate but must remain perpendicular to each other.

Vector. A line used to represent a vector quantity.

Vector Quantity. A quantity having magnitude and direction, e.g. force, distance, velocity, acceleration, momentum.

Addition of Vectors. Vectors are added by joining one to another and measuring the distance between their ends.

Subtraction of Vectors. Vectors are drawn from the same origin and the distance between their ends represents difference or change.

FRICTION

Force of Friction is the resistance to the motion between sliding surfaces.

$$F = \mu N$$

F = frictional resistance	(N)
μ = coefficient of friction	(—)
N = normal reaction	(N)

Resultant Reaction is the force applied by a plane surface to a moving body.

$$R = \sqrt{F^2 + N^2}$$

R = resultant reaction	(N)
F = frictional resistance	(N)
N = normal reaction	(N)

Friction Angle ϕ is the angle between R and N

$$\mu = \tan \phi$$

μ = coefficient of friction	(—)
$\tan \phi$ = tangent of the friction angle	(—)

KINEMATICS

Linear Kinematics relates to the motion of a body in a straight line.

Displacement is the distance moved by the body.

Velocity is the rate at which the body is moved.

$$s = vt$$

s = displacement	(m)
v = velocity	(m/s)
t = time taken	(s)

Variable Velocity (non uniform)

$$s = v_A t$$

v_A = average velocity	(m/s)

Uniform Change of Velocity

$$v_A = \frac{v_2 + v_1}{2}$$

v_2 = final velocity	(m/s)
v_1 = initial velocity	(m/s)

Displacement during Uniform Velocity Change

$$s = \left(\frac{v_2 + v_1}{2}\right) t$$

s = distance travelled	(m)
v_2 = final velocity	(m/s)
v_1 = initial velocity	(m/s)
t = time taken	(s)

Acceleration is the rate of change of velocity.

$$a = \frac{v_2 - v_1}{t}$$

a = uniform acceleration	(m/s^2)
v_2 = final velocity	(m/s)
v_1 = initial velocity	(m/s)
t = time taken	(s)

Derived Linear Kinematic Formulae

$$s = v_1 t + \tfrac{1}{2}at^2$$
$$v_2^2 = v_1^2 + 2as$$

s = distance travelled	(m)
v_1 = initial velocity	(m/s)
v_2 = final velocity	(m/s)
a = acceleration	(m/s^2)
t = time taken	(s)

Angular Kinematics relates to the motion of a revolving body.

Angular Displacement

$$s = \theta r$$

s = linear displacement	(m)
θ = angular displacement	(rad)
r = radius of curved path	(m)

Angular Velocity

$$v = \omega r$$

v = linear velocity	(m/s)
ω = angular velocity	(rad/s)
r = radius	(m)

Angular Acceleration

$$a = \alpha r$$

a = linear acceleration	(m/s^2)
α = angular acceleration	(rad/s^2)
r = radius	(m)

Derived Angular Kinematic Formulae

$$\theta = \left(\frac{\omega_2 + \omega_1}{2}\right)t$$
$$\alpha = \frac{\omega_2 - \omega_1}{t}$$
$$\theta = \omega_1 t + \tfrac{1}{2}\alpha t^2$$
$$\omega_2^2 = \omega_1^2 + 2\alpha\theta$$

θ = angular displacement	(rad)
ω_2 = final angular velocity	(rad/s)
ω_1 = initial angular velocity	(rad/s)
α = angular acceleration	(rad/s^2)
t = time	(s)

Projectiles. For bodies in free flight it is assumed that—
Horizontal velocity remains constant—air resistance is neglected.
Vertical velocity subject to constant change g = 9.81m/s^2.
When motion is downward g is positive, when upward g is negative.
Vertical velocity is zero at top of flight.
For flights above a given level, total time of flight = 2 × time to top of flight.

7

STRENGTH OF MATERIALS

Strength of Materials examines the behaviour of materials under direct or shear loading.

Direct Load is force applied axially in tension or compression.

Shear Load is force applied tangentially across a material.

Stress is a measure of materials resistance to load.

$$\sigma = \frac{F}{A}$$

σ = direct stress	(N/m^2)
F = load in tension or compression	(N)
A = area supporting load	(m^2)

$$\tau = \frac{F}{A}$$

τ = shear stress	(N/m^2)
F = shear load	(N)
A = area supporting load	(m^2)

Strain is a measure of the deformation of material under load.

$$\varepsilon = \frac{x}{l}$$

ε = direct strain	(—)
x = change of length	(m)
l = original length	(m)

Shear Strain is denoted by γ and has no units.

Modulus of Elasticity. A constant for any given material.

$$E = \frac{\sigma}{\varepsilon}$$

E = Young's modulus of elasticity	(N/m^2)
σ = direct stress	(N/m^2)
ε = direct strain	(—)

Modulus of Rigidity is a similar constant for conditions of shear.

$$G = \frac{\tau}{\gamma}$$

G = modulus of rigidity	(N/m^2)
τ = shear stress	(N/m^2)
γ = shear strain	(—)

Change of Length due to Change of Temperature

$$x = \alpha l \delta\theta$$

x = change of length	(m)
α = coefficient of linear expansion	(/°C)
l = original length	(m)
$\delta\theta$ = change of temperature	(°C)

BENDING OF BEAMS

Simply Supported Beam has reactions which can only give vertical support.

Cantilever has a single support at one end capable of applying a fixing moment.

Shear Force (SF) is the nett sum of the forces to one side of any point along the beam.

SF Diagram shows the shear force at any point on the beam.

Bending moment (BM) is the nett sum of the bending moments to one side of any point along the beam.

BM Diagram shows the bending moment at any point on the beam.

Points of Maximum BM occur at points of zero SF.

STRESS IN BEAMS

Beam Failure occurs at the point of maximum BM when the limiting bending stress of the beam material is exceeded.

Beam Equation

$$\frac{M}{I} = \frac{\sigma}{y} = \frac{E}{r}$$

M = bending moment	(Nm)
I = 2nd moment of area	(m^4)
σ = bending stress	(N/m^2)
y = distance from neutral axis	(m)
E = modulus of elasticity	(N/m^2)
r = radius of curvature	(m)

I Value for rectangles.

$$I = \frac{BD^3}{12}$$

I = 2nd moment of area	(m^4)
B = breadth of beam	(m)
D = depth of beam	(m)

I Value for round bars.

$$I = \frac{\pi D^4}{64}$$

I = 2nd moment of area	(m^4)
D = diameter of round bar	(m)

For hollow sections subtract the I value for the material removed.

TORSION

Torque is the turning moment applied to a twisting shaft.

$$T = Fr$$

T = applied torque	(Nm)
F = applied force	(N)
r = radius from axis of shaft	(m)

Torsion Equation

$$\frac{T}{J} = \frac{\tau}{r} = \frac{G\theta}{l}$$

T = applied torque	(Nm)
J = polar 2nd moment of area	(m^4)
τ = shear stress	(N/m^2)
r = radius from neutral axis	(m)
G = modulus of rigidity	(N/m^2)
θ = angle of twist	(rad)
l = length of shaft	(m)

J Value for solid shafts.

$$J = \frac{\pi D^4}{32}$$

J = polar 2nd moment of area	(m^4)
D = diameter of shaft	(m)

For a hollow shaft subtract the J value for the material removed.

Torsion Bars

$$W = \frac{T\theta}{2}$$

W = work done	(J)
T = applied torque	(Nm)
θ = angle of twist	(rad)

Transmission Shaft

$$W = T\theta$$

W = work done	(J)
T = applied torque	(Nm)
θ = angular displacement	(rad)

$$P = T\omega$$

P = power transmitted	(W)
T = applied torque	(Nm)
ω = angular velocity	(rad/s)

$$\omega = \frac{2\pi n}{60}$$

ω = angular velocity	(rad/s)
n = speed of rotation	(rev/min)

Couplings transmit the same power and torque as the shafts they connect.

$$T = \tau A N r$$

T = applied torque	(Nm)
τ = shear stress on bolts	(N/m^2)
A = area of section of bolt	(m^2)
N = number of bolts	(—)
r = radius of pitch circle	(m)

$$P = \frac{2\pi n T}{60}$$

P = power transmitted	(W)
n = speed of rotation	(rev/min)
T = applied torque	(Nm)

DYNAMICS

Momentum is a property of moving bodies given by the product.

$$\text{momentum} = \frac{m}{v}$$

m = mass of body	(kg)
v = velocity of body	(m/s)

Conservation of Momentum. During any change, unaffected by external force, the total momentum remains constant.

Momentum before = Momentum after

Conservation of Energy. Energy can be neither created nor destroyed. For any given process

Total energy before = Total energy after

Energy appears in many forms, e.g. as work, potential energy, kinetic energy, etc.

$$W = Fx$$

W = work	(J)
F = applied force	(N)
x = distance moved	(m)

$$PE = Wh$$

PE = potential energy (J)
W = weight of body (mg) (N)
h = vertical distance moved (m)

$$KE = \frac{mv^2}{2}$$

KE = kinetic energy (J)
m = mass of body (kg)
v = velocity of body (m/s)

Kinetic Energy of Rotation is stored in a revolving flywheel.

$$KE = \frac{m\omega^2 k^2}{2}$$

KE = kinetic energy of rotation (J)
m = mass of flywheel (kg)
ω = angular velocity (rad/s)
k = radius of gyration (m)

Work Done on Springs

$$W = \frac{Fx}{2}$$

W = work done (J)
F = applied force (N)
x = distance moved (m)

Centrifugal Force is the force trying to straighten the course of a body moving around a curved path.

$$CF = m\omega^2 r$$

m = mass of body (kg)
ω = angular velocity (rad/s)
r = radius of curved path (m)

MACHINES

Mechanical Advantage

$$MA = \frac{L}{E}$$

MA = mechanical advantage (—)
L = load overcome (N)
E = effort applied (N)

Velocity Ratio

$$VR = \frac{\text{distance moved by effort}}{\text{distance moved by load}}$$ (m) (m)

Efficiency

$$\eta = \frac{MA}{VR}$$

η = efficiency of machine (—)
MA = mechanical advantage (—)
VR = velocity ratio (—)

$$\eta = \frac{\text{work output}}{\text{work input}}$$

Velocity ratio of rope blocks = number of ropes supporting load.

$$\text{Velocity ratio of gear trains} = \frac{\text{number of teeth on follower}}{\text{number of teeth on driver}}$$

$$\text{Velocity ratio of hydraulic jacks} = \frac{\text{area of load piston}}{\text{area of effort piston}}$$

11

Law of Machine

$$E = aW + b$$

E = effort required	(N)
W = load overcome	(N)
a = a constant	(—)
b = a constant	(—)

HYDROSTATICS

Archimedes Principle states that when a body is partially or totally immersed in a fluid it is acted upon by an upthrust equal to the weight of fluid displaced.

Floating Bodies

Weight of body　　　　= weight of fluid displaced

Immersed volume of body = volume of fluid displaced

Sinking Bodies

Upthrust on body　　　= weight of fluid displaced

Total volume of body = volume of fluid displaced

Pressure Due to Head

$$p = h\rho g$$

p = pressure at depth h	(N/m^2)
h = head or depth of fluid	(m)
ρ = density of fluid	(kg/m^3)
g = 9·81	(m/s^2)

Load on an Immersed Surface

$$F = HA\rho g$$

F = hydrostatic load	(N)
H = depth to centroid from free surface	(m)
A = area of immersed surface	(m^2)
ρ = density of fluid	(kg/m^3)
g = 9·81	(m/s^2)

Centre of Pressure is the point at which the total hydrostatic load may be considered concentrated.

For a rectangle the centre of pressure is at two-thirds total depth.

HYDRAULICS

Flow Through Pipe

$$V = vA$$

V = volumetric rate of flow	(m^3/s)
v = velocity of fluid	(m/s)
A = area of pipe section	(m^2)

Flow Through an Orifice

$$v = \sqrt{2gh}$$

v = theoretical velocity of flow	(m/s)
h = head of fluid	(m)
g = 9·81	(m/s^2)

Coefficient of Velocity

$$C_V = \frac{\text{actual velocity of flow}}{\text{theoretical velocity of flow}}$$

Coefficient of Area

$$C_A = \frac{\text{area of jet}}{\text{area of orifice}}$$

Coefficient of Discharge

$$C_D = \frac{\text{actual rate of flow}}{\text{theoretical rate of flow}}$$

$$C_D = C_A \times C_V$$

Pumping

$F = pA$
F = force on piston (N)
p = fluid pressure (N/m²)
A = area of piston (m²)

$W = pV$
W = work done (J)
p = fluid pressure (N/m²)
V = volume discharged (m³)

$P = pV$
P = power required (W)
p = fluid pressure (N/m²)
V = volumetric rate of flow (m³/s)

Worked Examples

STATICS

Q.1 Four coplanar forces A, B, C and D act radially outwards from a common point with magnitudes of 80N, 50N, 100N and 150N respectively. The angular displacements of B, C and D reading clockwise from A are 60°, 150° and 300° respectively. Determine the magnitude and direction of the single additional force required for static equilibrium of the system.

A.

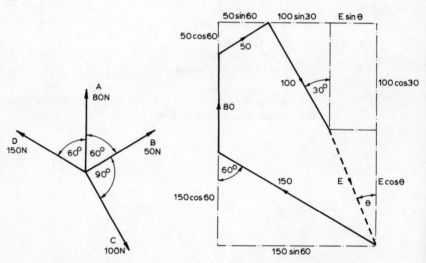

Fig. 1a   Space diagram

Fig. 1b   Vector diagram

The components of each force in the vector diagram (Fig. 1b) form a rectangle the opposite sides of which are equal in length. Each side represents the nett sum of all components of force in one direction.

Equating the horizontal components

$$150 \sin 60 = 50 \sin 60 + 100 \sin 30 + E \sin \theta$$

$$E \sin \theta = 150 \sin 60 - 50 \sin 60 - 100 \sin 30$$

$$= 129 \cdot 9 - 43 \cdot 3 - 50$$

$$= 36 \cdot 6 \tag{i}$$

Equating the vertical components

$$E \cos \theta + 100 \cos 30 = 150 \cos 60 + 80 + 50 \cos 60$$

$$E \cos \theta = 150 \cos 60 + 80 + 50 \cos 60 - 100 \cos 30$$

$$= 75 + 80 + 25 - 86 \cdot 6$$

$$= 93 \cdot 4 \tag{ii}$$

Dividing (i) by (ii)

$$\frac{E \sin \theta}{E \cos \theta} = \frac{36 \cdot 6}{93 \cdot 4}$$

$$\tan \theta = 0 \cdot 3919$$

$$\therefore \quad \theta = 21° \ 24'$$

Substituting into (i)

$$E \sin \theta = 36 \cdot 6$$

$$E = \frac{36 \cdot 6}{0 \cdot 3649}$$

$$= 100 \cdot 3$$

Force required for equilibrium = 100·3N at 158° 36′ from *A*.

Q.2   Four coplanar forces A, B, C and D act outwards from a point which is in equilibrium. Force A is 3kN and force C is 2kN. Forces B, C and D act at 60°, 135° and 225° respectively measured clockwise from A. Determine the magnitude of forces B and D by calculation and by graphical means.

A.

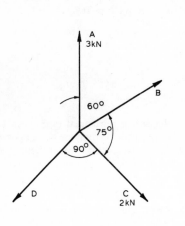

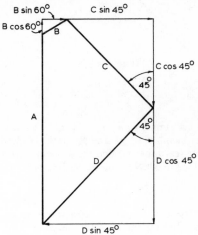

Fig. 2a   Space diagram   Fig. 2b   Vector diagram

15

Equating for equilibrium the components of forces shown in Fig. 2b horizontally

$$B \sin 60 + 2 \sin 45 = D \sin 45 \qquad \text{(i)}$$

vertically

$$B \cos 60 + 3 = D \cos 45 + 2 \cos 45 \qquad \text{(ii)}$$

Subtracting (ii) from (i)

$$B \sin 60 + 1.414 - B \cos 60 - 3 = -1.414$$

$$B(\sin 60 - \cos 60) = -1.414 - 1.414 + 3$$

$$B = \frac{3 - 2.828}{0.866 - 0.5}$$

$$= \frac{0.172}{0.366}$$

$$= 0.47 \text{kN}$$

Substituting into (i)

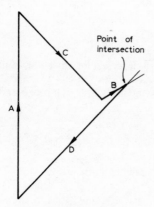

$$D \sin 45 = 0.47 \times 0.866 + 1.414$$

$$D = \frac{0.407 + 1.414}{0.7071}$$

$$= 2.58 \text{kN}$$

The vector diagram Fig. 2c shows the known forces A and C added vectorially. The diagram is *closed* for equilibrium by adding vectors B and D in their respective directions. The magnitude of B and D is found by measurement to the point of intersection

$$\text{Force B} = 0.5 \text{kN}$$

$$\text{Force D} = 2.6 \text{kN}$$

Fig. 2c   Vector diagram

Q.3   A simple roof truss consists of a simply supported horizontal beam with two other members at 50° and 40° to the horizontal respectively. Determine the magnitude of the reactions and find the force in each member, stating the nature of these forces, if a load of 60kN is applied to the apex of the structure.

A.   The space diagram, Fig. 3a, shows the arrangement of the members forming the roof truss. The magnitude of the forces in these members is determined from the vector diagram Fig. 3b. Triangle ABC represents the three forces at the apex of the structure, triangle DAC represents the forces at R_2 and triangle BDC represents the forces at R_1.

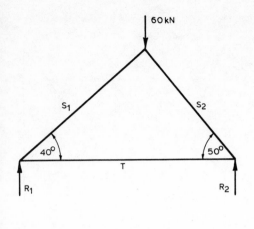

Fig. 3a   Space diagram

Fig. 3b   Vector diagram

By applying the basic trigonometrical ratios

$$AC = 60 \cos 40°$$

$\therefore$ Compressive force in $S_2 = 46\text{kN}$

$$BC = 60 \sin 40°$$

$\therefore$ Compressive force in $S_1 = 38\cdot6\text{kN}$

$$DC = 46 \sin 40°$$

$\therefore$ Tensile force in T $= 29\cdot6\text{kN}$

$$BD = 29\cdot6 \tan 40°$$

$\therefore$ Reaction R_1 $= 24\cdot8\text{kN}$

$$DA = 29\cdot6 \tan 50°$$

$\therefore$ Reaction R_2 $= 35\cdot2\text{kN}$

Q.4   A beam of non-uniform section is lifted by two slings of unequal length attached at its extreme ends. The beam is 5m long and its centre of gravity is 1m from the right hand end. The slings are fixed to a single hook which is raised to a point 5m above the ground, the beam then makes an angle of 30° to the ground with its left hand end 500mm from the ground. If the mass of the beam is 15Mg determine the tension in each sling.

A.

$$\text{Weight of beam } W = 15\,000 \times 9\cdot81$$
$$= 147\text{kN}$$

By measurement, (Fig. 4b)

$$\text{Tension in sling } S_1 = 66\cdot5\text{kN}$$

$$\text{Tension in sling } S_2 = 103\text{kN}$$

17

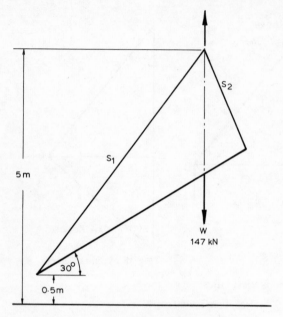

Fig. 4a   Space diagram   Fig. 4b   Vector diagram

Q.5 A derrick 12m long is supported by a single stay. The angle between the derrick post and the derrick is 75° whilst the included angle between the stay and the derrick is 25°. The derrick supports a load of 1750kg and its own mass is 500kg.

Calculate:
(a)   The force in the supporting stay.
(b)   The compressive force in the derrick.

A.

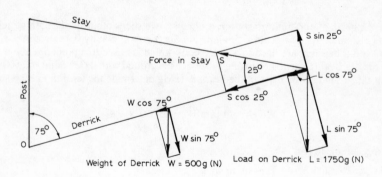

Fig. 5   Derrick loading

By resolving all forces perpendicular and parallel to the derrick and taking moments about 0

$$S \sin 25° \times 12 = L \sin 75° \times 12 + W \sin 75° \times 6$$

Dividing by 12

$$S \sin 25° = L \sin 75° + 0.5W \sin 75°$$

$$S = \frac{(L + 0.5W) \sin 75°}{\sin 25°}$$

$$= \frac{(17\,170 + 2453)0.966}{0.423}$$

$$= 44.8 \times 10^3 \text{N}$$

The compressive force on the foot of the derrick is given by the nett sum of the forces acting along the axis of the derrick

$$\text{Compressive force} = W \cos 75° + L \cos 75° + S \cos 25°$$

$$= 4905 \times 0.259 + 17\,170 \times 0.259 + 44\,800 \times 0.906$$

$$= 1270 + 4444 + 40\,600$$

$$= 46\,314 \text{ N}$$

The supporting force in stay $= 44.8\text{kN}$

The compressive force in derrick $= 46.3\text{kN}$

Q.6 A casting weighing 981N is lifted by two ropes as shown. Find the force in the ropes when the casting is just clear of the ground. If the length of the casting AB = 4m determine, by calculation, the position of its centre of gravity.

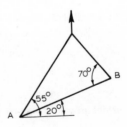

A.

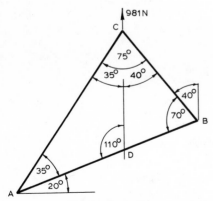

Fig. 6a   Space diagram

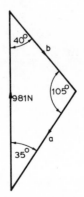

Fig. 6b   Vector diagram

The forces in the ropes AC and BC are given by a and b respectively in the vector diagram Fig. 6b using the Sine Rule

$$\frac{a}{\sin 40°} = \frac{b}{\sin 35°} = \frac{981}{\sin 105°}$$

now $\sin 105° = \sin 75°$

$$\therefore \quad a = \frac{981 \times \sin 40°}{\sin 75°}$$

$$= \frac{981 \times 0.643}{0.966}$$

$$= 653N$$

and

$$b = \frac{981 \times \sin 35°}{\sin 75°}$$

$$= \frac{981 \times 0.574}{0.966}$$

$$= 583N$$

Applying the Sine Rule to triangle ABC in Fig. 6a

$$\frac{AC}{\sin 70°} = \frac{AB}{\sin 75°} \qquad \text{where } AB = 4m$$

$$\therefore \quad AC = \frac{4 \sin 70°}{\sin 75°}$$

For equilibrium the centre of gravity D must be vertically below the point of suspension C.

In triangle ADC

$$\frac{AD}{\sin 35°} = \frac{AC}{\sin 110°}$$

Substituting AC gives

$$AD = \frac{4 \sin 70°}{\sin 75°} \times \frac{\sin 35°}{\sin 110°}$$

now $\sin 110° = \sin 70°$

$$\therefore \quad AD = \frac{4 \times 0.574}{0.966}$$

$$= 2.38m$$

The force in the ropes $= 653N$ and $583N$

The position of C.G. from A $= 2.38m$

Q.7 In an internal combustion engine the diameter of the piston is 375mm, the stroke is 0·75m and the length of the connecting rod is 1·5m. When the crank is 30° past top dead centre the mean cylinder pressure is 3·2MN/m² calculate:
  (a)   The force in the connecting rod.
  (b)   The torque applied to the crankshaft.

A.

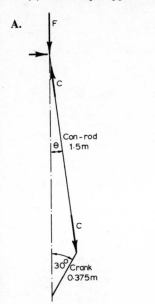

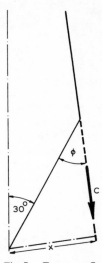

Fig. 7a   Space diagram   Fig. 7b   Vector diagram   Fig. 7c   Torque = Cx

Using the Sine Rule (Fig. 7a)

$$\sin \theta = \frac{0·375 \sin 30°}{1·5}$$

$$= 0·125$$

$$\therefore \quad \theta = 7° \, 11'$$

The force on the piston
$$F = pA$$

$$= 3·2 \times 10^6 \times \frac{\pi}{4} \times 0·375^2$$

$$= 0·354 \times 10^6 \text{N}$$

The forces on the crosshead are in equilibrium (Fig. 7b) hence the force on the con-rod

$$C = \frac{F}{\cos 7° \, 11'}$$

$$= \frac{0·354 \times 10^6}{0·99}$$

$$= 0·357 \text{MN}$$

21

The angle between the crank and the con-rod

$$= 180° - 30° - 7° 11'$$
$$= 142° 49'$$

The torque on the crank (Fig. 7c) $= Cx$

where

$$x = 0.375 \sin \phi$$

and

$$\phi = 180° - 142° 49'$$

$\therefore$ Torque on crankshaft

$$= 0.354 \times 10^6 \times 0.375 \sin 37° 11'$$
$$= 0.0802 \times 10^6 \text{Nm}$$

Force in the connecting rod $\quad = 357\text{kN}$

Torque on the crankshaft $\quad\quad = 80.2\text{kNm}$

Q.8 A uniform ladder of mass 200kg is hinged at deck level and hangs down parallel to the ships side at 60° to the horizontal. A steel chain attached to the bottom of the ladder leads up at 40° to the vertical to a winch on the same level as the hinge. Calculate the tension in the chain and the reaction at the hinge.

A.

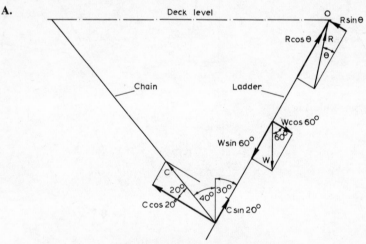

Fig. 8   Components of forces on ladder

In Fig. 8 let

$l =$ length of ladder

$W =$ weight of ladder

$R =$ reaction at the hinge

$C =$ tension in the chain.

Fig. 8 shows all the forces resolved into components perpendicular and parallel to the axis of the ladder.

By equating the moments of force about the hinge O

$$W \cos 60° \times \frac{l}{2} = C \cos 20° \times l$$

$$\therefore \quad C = \frac{W \cos 60°}{2 \cos 20°}$$

$$= \frac{200 \times 9 \cdot 81 \times 0 \cdot 5}{2 \times 0 \cdot 94}$$

$$= 522\text{N}$$

Equating the forces perpendicular to the ladder

$$R \sin \theta + C \cos 20° = W \cos 60°$$

$$\therefore \quad R \sin \theta = 200 \times 9 \cdot 81 \times 0 \cdot 5 - 522 \times 0 \cdot 94$$

$$= 981 - 490 \cdot 5$$

$$= 490 \cdot 5 \qquad \text{(i)}$$

Equating the forces parallel to the plane of the ladder

$$R \cos \theta + C \sin 20° = W \sin 60°$$

$$\therefore \quad R \cos \theta = 200 \times 9 \cdot 81 \times 0 \cdot 866 - 522 \times 0 \cdot 342$$

$$= 1700 - 179$$

$$= 1521 \qquad \text{(ii)}$$

Dividing (i) by (ii)

$$\frac{R \sin \theta}{R \cos \theta} = \frac{490 \cdot 5}{1521}$$

$$\therefore \quad \tan \theta = 0 \cdot 322$$

and

$$\theta = 17° \; 50'$$

Substituting into (i)

$$R = \frac{490 \cdot 5}{\sin 17° \; 50'}$$

$$= 1600\text{N}$$

The tension in the chain $= 522\text{N}$

and the reaction at the hinge $= 1600\text{N}$ acting at $17° \; 50'$ to the axis of the ladder.

Q.9 A uniform beam AB has a single support 1·7m from A. The mass required to balance the beam is 100kg applied at A. If the point of suspension is moved to a point 2m from A and the mass required to balance the beam is reduced to 50kg find the length and the mass of the beam.

A.

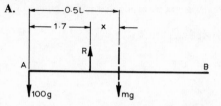

Fig. 9a   Initial loading

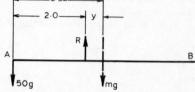

Fig. 9b   Final loading

Let the length of the beam $= L$
and the mass of the beam $= m$
The weight of the beam $= mg$
and the applied force in Fig. 9a $= 100 \times g$
and in Fig. 9b $= 50 \times g$

Equating moments of force about the point of suspension.

For Fig. 9a

$$100g \times 1 \cdot 7 = mg \times x$$

$$\text{where} \quad x = 0 \cdot 5L - 1 \cdot 7$$

$$\therefore \; m = \frac{100 \times 1 \cdot 7}{(0 \cdot 5L - 1 \cdot 7)} \qquad \text{(i)}$$

For Fig. 9b

$$50g \times 2 = mg \times y$$

$$\text{where} \quad y = 0 \cdot 5L - 2$$

$$\therefore \; m = \frac{50 \times 2}{(0 \cdot 5L - 2)} \qquad \text{(ii)}$$

Substituting (i) into (ii)

$$\frac{170}{(0 \cdot 5L - 1 \cdot 7)} = \frac{100}{(0 \cdot 5L - 2)}$$

$$170(0 \cdot 5L - 2) = 100(0 \cdot 5L - 1 \cdot 7)$$

$$85L - 340 = 50L - 170$$

$$85L - 50L = 340 - 170$$

$$L = \frac{170}{35}$$

$$= 4 \cdot 86 \text{m}$$

Substituting into (ii)

$$m = \frac{100}{(0 \cdot 5 \times 4 \cdot 86 - 2)}$$

$$= \frac{100}{(2 \cdot 43 - 2)}$$

$$= 232 \cdot 7 \text{kg}$$

Length of beam $= 4 \cdot 86$m
Mass of beam $= 232 \cdot 7$kg

FRICTION

Q.10 A weight rests on a rough inclined plane which is set at an angle of 10° to the horizontal. The coefficient of friction between the weight and the plane is 0·25. If a horizontal force of 100N will just move the body down the plane, calculate the magnitude of the weight.

A.

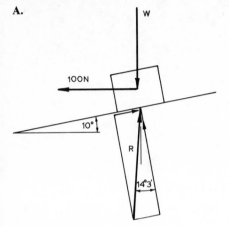

Fig. 10a   Space diagram   Fig. 10b   Vector diagram

Since the tangent of the friction angle ϕ is equal to the coefficient of friction

$$\tan \phi = 0·25$$

$$\therefore \quad \phi = 14° \, 3'$$

The resultant reaction R acts at 4° 3′ to the vertical. From the vector diagram Fig. 10b

$$W = \frac{100}{\tan 4° \, 3'}$$

$$= \frac{100}{0·0707}$$

$$= 1418N$$

Weight of body $= 1418N$

Q.11 The effort required to just stop a body from sliding down an incline is 9·58 % of its weight and the force required to pull it up the incline is 71·33 % of its weight. Calculate the angle of the plane and the coefficient of friction if the effort and force are applied parallel to the incline.

A.   Equating the forces parallel to the plane in Fig. 11a

$$F + 0·0958W = W \sin \theta$$

$$\therefore \quad F = W \sin \theta - 0·0958W \qquad \text{(i)}$$

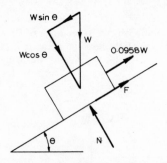

Fig. 11a Holding on incline

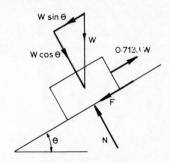

Fig. 11b Pulling up incline

and in Fig. 11b

$$F + W \sin \theta = 0.7133W$$

$$\therefore \quad F = 0.7133W - W \sin \theta \qquad \text{(ii)}$$

Equating forces perpendicular to the plane in both diagrams

$$N = W \cos \theta \qquad \text{(iii)}$$

Substituting the above values into the friction formula

$$F = \mu N$$

using equations (i) and (iii)

$$W \sin \theta - 0.0958W = \mu W \cos \theta \qquad \text{(iv)}$$

and equations (ii) and (iii)

$$0.7133W - W \sin \theta = \mu W \cos \theta \qquad \text{(v)}$$

Since $\mu W \cos \theta$ is common to both the resulting equations then equating (iv) and (v)

$$W \sin \theta - 0.0958W = 0.7133W - W \sin \theta$$

Dividing by W gives

$$2 \sin \theta = 0.7133 + 0.0958$$

$$\sin \theta = \frac{0.8091}{2}$$

$$= 0.4046$$

$$\therefore \quad \theta = 23° \, 52'$$

Substituting θ into equation (iv) gives

$$W \sin 23° \, 52' - 0.0958W = \mu W \cos 23° \, 52'$$

hence

$$\mu = \frac{W \times 0.4046 - 0.0958W}{W \times 0.9145}$$

26

Again dividing by W gives

$$\mu = \frac{0.4046 - 0.0958}{0.9145}$$

$$= \frac{0.3088}{0.9145}$$

$$= 0.338$$

$$\text{Angle of incline} = 23° 52'$$

$$\text{Coefficient of friction} = 0.338$$

Q.12 A sliding door hangs from a horizontal rail on two wheels. The wheels are equidistant from the centre of gravity of the door and are 1m apart. One wheel is seized and has a coefficient of friction of 0.2, the other wheel is free to rotate and the door handle is 1m below the seized wheel. Calculate the forces required at the handle to open and close the door if its weight is 250N.

A.

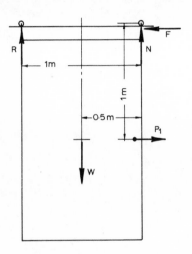

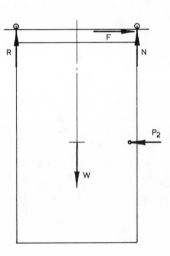

Fig. 12a   Moving door to right   Fig. 12b   Moving door to left

Referring to Fig. 12a and applying the conditions of equilibrium to the horizontal forces

$$F = P_1$$

to the vertical forces

$$N + R = W$$

and to the moments of force taken about the seized wheel

$$(R \times 1) = (W \times 0.5) + (P_1 \times 1)$$

Substituting these equations into the friction formula

$$F = \mu N$$

gives

$$P_1 = \mu(W - R)$$
$$= \mu W - \mu(0 \cdot 5W + P_1)$$
$$P_1 + \mu P_1 = \mu W - 0 \cdot 5 \mu W$$
$$P_1(1 + \mu) = 0 \cdot 5 \mu W$$
$$P_1 = \frac{0 \cdot 5 \times 0 \cdot 2 \times 250}{1 + 0 \cdot 2}$$
$$= 20 \cdot 8\text{N}$$

When the door is moved in the opposite direction, Fig. 12b

$$F = P_2$$

and

$$N + R = W$$

Equating moments

$$(R \times 1) = (W \times 0 \cdot 5) - (P_2 \times 1)$$

Substituting into

$$F = \mu N$$

gives

$$P_2 = \mu(W - R)$$
$$= \mu W - \mu(0 \cdot 5W - P_2)$$
$$P_2 - \mu P_2 = \mu W - 0 \cdot 5 \mu W$$
$$P_2(1 - \mu) = 0 \cdot 5 \mu W$$
$$P_2 = \frac{0 \cdot 5 \times 0 \cdot 2 \times 250}{1 - 0 \cdot 2}$$
$$= 31 \cdot 3\text{N}$$

Forces to open and close door = 20·8N and 31·3N.

Q.13 A uniform ladder 10m long has a mass of 10kg and rests against a vertical wall. The coefficients of friction between the ladder and the wall and the ladder and the ground are 0·4 and 0·7 respectively. Calculate the angle at which the ladder must be placed to the ground so that a man of 60kg mass can climb two-thirds of the ladder without it slipping.

A. Applying the friction formula to the points of contact of the ladder to the wall

$$F_1 = 0 \cdot 4N_1 \qquad \text{(i)}$$

and the ladder to the ground

$$F_2 = 0 \cdot 7N_2 \qquad \text{(ii)}$$

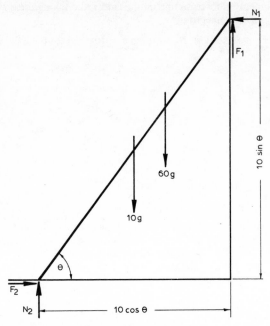

Fig. 13   Forces on ladder

Equating for equilibrium vertical forces

$$N_2 + F_1 = 10g + 60g \qquad \text{(iii)}$$

and horizontal forces

$$N_1 = F_2$$

Substituting equation (ii) $N_1 = 0 \cdot 7 N_2$

then equation (iii) $= 0 \cdot 7 (70g - F_1)$

and finally equation (i) $= 0 \cdot 7 (70g - 0 \cdot 4 N_1)$

gives

$$N_1 = 49g - 0 \cdot 28 N_1$$

$$\therefore \quad N_1 + 0 \cdot 28 N_1 = 49g$$

$$N_1 = \frac{49 \times 9 \cdot 81}{1 \cdot 28}$$

$$= 376 \text{N}$$

Substituting into equation (i)

$$F_1 = 0 \cdot 4 \times 376$$

$$= 150 \text{N}$$

29

Equating the moments of forces acting on the ladder and taking these moments about the point of contact with the ground

$$F_1 \times 10 \cos\theta + N_1 \times 10 \sin\theta = 10g \times 5 \cos\theta + 60g \times 6.67 \cos\theta$$

Dividing by $\cos\theta$

$$150 \times 10 + 3760 \tan\theta = 50g + 400g$$

$$3760 \tan\theta = 450 \times 9.81 - 1500$$

$$\therefore \quad \tan\theta = \frac{4415 - 1500}{3760}$$

$$= 0.775$$

$$\therefore \quad \theta = 37.8°$$

Limiting angle of ladder $= 37° \ 48'$

KINEMATICS

Q.14 A launch accelerates uniformly from standstill to a constant speed of 10 knots in 600 seconds. After covering 3.7km at constant speed it retards uniformly to rest. The total distance covered is 7.1km. Take 1 knot as equal to 1.85km/hour and determine:
(a)   The distance travelled during acceleration,
(b)   The total voyage time.

A.

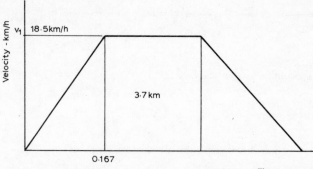

Fig. 14   Velocity-time diagram

$$\text{Maximum velocity of launch } v = 1.85 \times 10$$

$$= 18.5 \text{km/h}$$

$$\text{Time taken for acceleration} = \frac{600}{3600}$$

$$= 0.167 \text{h}$$

Distance travelled during acceleration from standstill,

$$s = \frac{18.5}{2} \times 0.167$$

$$= 1.54 \text{km}$$

$$\text{Time taken at constant velocity } t = \frac{3\cdot7}{18\cdot5}$$

$$= 0\cdot2\text{h}$$

Distance travelled while decelerating to standstill,

$$= 7\cdot1 - (1\cdot54 + 3\cdot7)$$

$$= 1\cdot86\text{km}$$

Time taken

$$= \frac{1\cdot86}{18\cdot5 \times 0\cdot5}$$

$$= 0\cdot201\text{h}$$

Total time for voyage

$$= 0\cdot167 + 0\cdot2 + 0\cdot201$$

$$= 0\cdot568\text{h}$$

Acceleration distance $= 1\cdot54\text{km}$

Time for entire voyage $= 0\cdot568\text{h}$

Q.15 The distance between two railway stations is 2·5km. A train starts from one station and accelerates at 1·2m/s² until it reaches a speed of 108km/hour. It continues to travel at constant speed for some time then finally retards at 0·8m/s² until it reaches the next station. Calculate the time taken for the entire journey.

A.

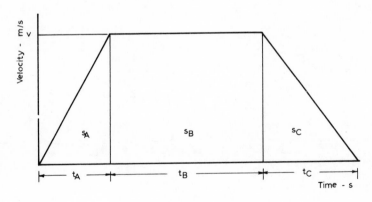

Fig. 15   Velocity-time diagram

Maximum velocity of train $v = \dfrac{108 \times 1000}{3600}$

$$= 30\text{m/s}$$

From

$$v_2 = v_1 + at$$

$$t = \frac{v_2 - v_1}{a}$$

$$\therefore \quad t_A = \frac{30 - 0}{1 \cdot 2}$$

$$= 25s$$

and

$$t_C = \frac{0 - 30}{-0 \cdot 8}$$

$$= 37 \cdot 5s$$

Using

$$v = \frac{s}{t}$$

$$t_B = \frac{s_B}{30} \qquad \text{(i)}$$

Since the total distance travelled $= s_A + s_B + s_C$

then

$$s_B = 2500 - s_A - s_C \qquad \text{(ii)}$$

and from

$$v_2^2 = v_1^2 + 2as$$

$$s = \frac{v_2^2 - v_1^2}{2a}$$

$$\therefore \quad s_A = \frac{30^2 - 0}{2 \times 1 \cdot 2}$$

$$= 375m$$

and

$$s_C = \frac{0 - 30^2}{2 \times (-0 \cdot 8)}$$

$$= 562 \cdot 5m$$

Substituting into (ii)

$$s_B = 2500 - 375 - 562 \cdot 5$$

$$= 1562 \cdot 5m$$

and into (i)

$$t_B = \frac{1562 \cdot 5}{30}$$

$$= 52 \cdot 1s$$

Total time taken for journey $= t_A + t_B + t_C$

$$= 25 + 52{\cdot}1 + 37{\cdot}5$$

$$= 114{\cdot}6\text{s}$$

Time taken for entire journey $= 114{\cdot}6\text{s}$

Q.16   A velocity-time graph is the shape of a circular arc which forms a segment of a circle with the time base. The length of the time base is 160mm which represents a time of 32 seconds, the maximum height of the arc is 21mm and this represents the maximum velocity of 10·5m/s. Calculate the displacement in metres represented by the graph.

A.

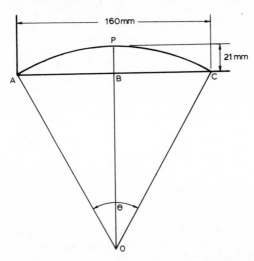

Fig. 16   Velocity-time graph showing the geometry of the circular arc

Let D be the diameter of a circle of which APC is an arc. By the theorem of crossed chords

$$\text{BP} \times (\text{D} - \text{BP}) = \text{AB} \times \text{BC}$$

$$\therefore\quad \text{D} = \frac{\text{AB} \times \text{BC}}{\text{BP}} + \text{BP}$$

$$= \frac{80 \times 80}{21} + 21$$

$$= 326\text{mm}$$

The radius of the arc OA $= \text{OC}$

$$= \frac{326}{2}$$

$$= 163\text{mm}$$

33

and the vertical OB $\quad = 163 - 21$

$$= 142\text{mm}$$

Since OBA is a right angle

$$\tan\frac{\theta}{2} = \frac{80}{142}$$

$$= 0.563$$

$$\frac{\theta}{2} = 29.4°$$

and

$$\theta = 58.8°$$

The area of the sector OAPC $= \dfrac{\pi}{4}D^2 \times \dfrac{\theta}{360}$

$$= \frac{\pi \times 326^2 \times 58.8}{4 \times 360}$$

$$= 13\,630\text{mm}^2 \qquad\qquad\qquad\text{(i)}$$

The area of triangle OAC $\quad = \dfrac{\text{OB} \times \text{AC}}{2}$

$$= \frac{160 \times 142}{2}$$

$$= 11\,360\text{mm}^2 \qquad\qquad\qquad\text{(ii)}$$

Area of segment (i) − (ii) $\quad = 13\,630 - 11\,360$

$$= 2270\text{mm}^2$$

Given that 160mm represents 32s then 1mm represents 0.2s, also that 21mm represents 10.5m/s then 1mm represents 0.5m/s. It follows that the product 1mm × 1mm, 1mm^2 represents (0.2s × 0.5m/s) 0.1m.

The area under the graph represents the total distance travelled

$$= 2270 \times 0.1$$

$$= 227\text{m}$$

Q.17 A balloon rises at a constant velocity of 10m/s. At a height of 100m the balloon bursts. Neglecting air resistance find the time taken for the balloon to return to the ground.

A. Let A be the ground level, B the 100m level and C be the maximum height of the flight of the balloon. At 100m and rising the balloon has an initial velocity $v_B = 10$m/s, and an acceleration $a = -9.81$m/s^2. At the top of flight, velocity $v_C = 0$.

Using $\qquad\qquad v_C = v_B + at$

$$t = \frac{v_C - v_B}{a}$$

$$= \frac{0 - 10}{-9.81}$$

$$= 1.02\text{s}$$

Total time of flight above 100m $= 2 \times 1\cdot02$

$$= 2\cdot04s$$

Falling from 100m

$$v_B = 10m/s$$
$$a = 9\cdot81m/s^2$$

and the distance

$$s_{BA} = 100m$$

Using $s_{BA} = v_B t + 0\cdot5at^2$ and substituting these values

$$4\cdot905t^2 + 10t - 100 = 0.$$

Using the general solution for quadratic equations

$$t = \frac{-b \pm \sqrt{b^2 - 4ac}}{2a}$$

$$t = \frac{-10 \pm \sqrt{10^2 + 4 \times 4\cdot905 \times 100}}{2 \times 4\cdot905}$$

$$= \frac{-10 \pm \sqrt{100 + 1962}}{9\cdot81}$$

$$= \frac{-10 \pm 45\cdot4}{9\cdot81}$$

Ignoring the negative value

$$t = \frac{35\cdot4}{9\cdot81}$$

Time of fall from 100m $\qquad = 3\cdot61s$

Total time of free flight $\qquad = 2\cdot04 + 3\cdot61$

$$= 5\cdot65s$$

Time of flight after bursting $\quad = 5\cdot65s$

Q.18 A body is allowed to fall from a certain height and travels through the final 450m in 7 seconds. Find:
  (a) The total distance travelled.
  (b) The velocity at the end of the flight.
  (c) The duration of the flight.

A. Let A be the point of release, B be the 450m level, and C the level of the ground.

Acceleration is gravitational $g = 9\cdot81m/s^2$

For the 450m flight between B and C, using

$$s = v_1 t + \tfrac{1}{2}gt^2$$

35

The initial velocity $\qquad v_B = \left(s - \dfrac{gt^2}{2}\right)\dfrac{1}{t}$

$$= \frac{s}{t} - \frac{gt}{2}$$

$$= \frac{450}{7} - \frac{9\cdot81 \times 7}{2}$$

$$= 64\cdot3 - 34\cdot3$$

$$= 30\text{m/s}$$

Using

$$v_2^2 = v_1^2 + 2gs$$

The distance $\qquad s_{AB} = \dfrac{v_B^2 - v_A^2}{2g}$

$$= \frac{30^2 - 0}{2 \times 9\cdot81}$$

$$= 45\cdot9\text{m}$$

The total distance travelled $\quad = 450 + 45\cdot9$

$$= 495\cdot9\text{m} \qquad\qquad (a)$$

Also using

$$v_2^2 = v_1^2 + 2gs$$

gives the final velocity

$$v_C^2 = v_B^2 + 2gs_{BC}$$

$$= 30^2 + 2 \times 9\cdot81 \times 450$$

$$= 900 + 8830$$

$$\therefore \quad v_C = \sqrt{9730}$$

$$= 98\cdot6\text{m/s} \qquad\qquad (b)$$

From

$$v_2 = v_1 + gt$$

The duration of flight $\qquad t = \dfrac{v_C - v_A}{g}$

$$= \frac{98\cdot6 - 0}{9\cdot81}$$

$$= 10\cdot06\text{s} \qquad\qquad (c)$$

Total distance travelled $\qquad = 495\cdot9\text{m}$

The final velocity $\qquad\qquad = 98\cdot6\text{m/s}.$

The duration of flight $\qquad = 10\cdot06s$

Q.19 A hammer is dropped from a height of 4·905m. Neglecting air resistance find the time taken to reach the ground. The hammer comes to rest in 13mm after impact. Find the average retardation in m/s².

A.

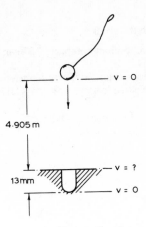

4·905 m

13mm

Fig. 19   Hammer imprint

Considering the period of free fall while the acceleration is gravitational

$$s = v_1t + \tfrac{1}{2}gt^2$$

Hence

$$t^2 = \frac{2s}{g} \qquad \text{since } v_1 = 0$$

$$\therefore \quad t = \sqrt{\frac{2 \times 4\cdot905}{9\cdot81}}$$

$$= 1\text{s}$$

Also

$$v_2 = v_1 + gt$$
$$v_2 = 0 + 9\cdot81 \times 1$$
$$= 9\cdot81\text{m/s}$$

This is the maximum velocity attained and will be the initial velocity during the deceleration after impact. The final velocity is zero.

     Thus using

$$v_2^2 = v_1^2 + 2as$$
$$0 = 9\cdot81^2 + 2a \times 0\cdot013$$

and

$$a = -\frac{9\cdot81^2}{2 \times 0\cdot013}$$

$$= -3700\text{m/s}^2$$

Time to reach the ground $\quad = 1\text{s}$

Average retardation after impact $= 3700\text{m/s}^2$

Q.20 A thin cord is wrapped around a 75mm diameter shaft. A mass is attached to the end of the cord. When released from rest the mass falls through 2·1m in 3 seconds. Find:

   (a)   The angular velocity of the shaft in rad/s.

   (b)   The angular acceleration in rad/s².

A.

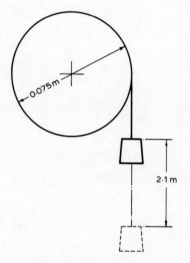

Fig. 20   Rotating shaft

Using

$$s = \left(\frac{v_1 + v_2}{2}\right)t$$

The final velocity of the mass, $v_2 = \dfrac{2s}{t} - v_1$

$$= \frac{2 \times 2\cdot1}{3} - 0$$

$$= 1\cdot4\text{m/s}$$

Since this represents the rate of displacement of the cord from the shaft

$$\omega = \frac{v_2}{r}$$

$$= \frac{1\cdot4}{0\cdot5 \times 0\cdot075}$$

$$= 37\cdot3\text{rad/s}$$

The acceleration of the mass $\quad a = \dfrac{v_2 - v_1}{t}$

$$= \frac{1\cdot4 - 0}{3}$$

$$= 0\cdot467\text{m/s}^2$$

This also represents the acceleration of the shaft surface

Hence angular acceleration $\quad \alpha = \dfrac{a}{r}$

$$= \frac{0\cdot467}{0\cdot5 \times 0\cdot075}$$

$$= 12\cdot4\text{rad/s}^2$$

Angular velocity of shaft $\qquad = 37\cdot3\text{rad/s}$

Angular acceleration of shaft $\quad = 12\cdot4\text{rad/s}^2$

Q.21 A projectile is fired from the ground at 25m/s and at 20° to the vertical. Find the horizontal distance covered and the height of the projectile after 3 seconds. Also find the distance travelled to the point of impact with the ground.

A.

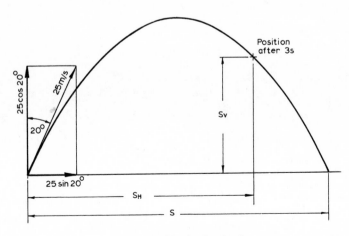

Fig. 21   Trajectory of projectile

The components of the initial velocity are: in the horizontal plane $v_H = 25 \sin 20°$, and in the vertical plane $v_V = 25 \cos 20°$.

The horizontal distance travelled in three seconds, assuming no air resistance

$$s_H = v_H t$$
$$= 25 \sin 20° \times 3$$
$$= 25·65\text{m}$$

The height of the projectile after three seconds with $g = -9·81\text{m/s}^2$ is found from

$$s_V = v_V t + \tfrac{1}{2}gt^2$$

Hence

$$s_V = 25 \cos 20° \times 3 - 0·5 \times 9·81 \times 3^2$$
$$= 70·5 - 44·2$$
$$= 26·3\text{m}$$

Using the same formula the time to impact with the ground where $s_V = 0$ is given by

$$0 = 25 \cos 20° \times t - 0·5 \times 9·81 \times t^2$$

Isolating the factor t

$$0 = t(25 \cos 20° - 0·5 \times 9·81t)$$

it follows that either

$$t = 0$$

or

$$0 = 25 \cos 20° - 0·5 \times 9·81t$$

From the latter

$$0·5 \times 9·81t = 25 \cos 20°$$
$$t = \frac{25 \cos 20°}{0·5 \times 9·81}$$
$$= 4·8\text{s}$$

Horizontal distance covered

$$s = v_H \times 4·8$$
$$= 25 \sin 20° \times 4·8$$
$$= 41\text{m}$$

In 3 seconds horizontal distance $= 25·65\text{m}$

and the vertical height $= 26·3\text{m}$

Horizontal distance to impact $= 41\text{m}$

Q.22 A tanker 100m long and a tug 20m long are both steaming on a parallel course at 10km/hour with the bow of the tug abreast of the tanker's stern. The tanker reduces speed at a rate of 0·05m/s² while the tug increases speed at a rate of 0·1m/s². Find:
(a) The time for their respective positions to be reversed.
(b) The speed of each vessel at this point.

A.

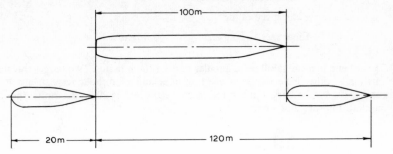

Fig. 22   Positions of tug and tanker

Consider the movement of the tug relative to the tanker. The distance moved by the tug

$$s = 120\text{m}$$

the initial relative velocity $v_1 = 0$

and the relative acceleration $a = 0\cdot1 + 0\cdot05$

$$= 0\cdot15\text{m/s}^2$$

Let the time to change positions be t, then using

$$s = v_1 t + \tfrac{1}{2}at^2$$

$$t^2 = \frac{2s}{a}$$

and

$$t = \sqrt{\frac{2 \times 120}{0\cdot15}}$$

$$= 40\text{s}$$

Since the absolute velocity of the tug and the tanker is

$$v_1 = \frac{10 \times 1000}{3600}$$

$$= 2\cdot78\text{m/s}$$

The actual velocity of the tug after 40 seconds is given by

$$v_2 = v_1 + at$$

$$= 2\cdot78 + 0\cdot1 \times 40$$

$$= 6\cdot78\text{m/s}$$

and the actual velocity of the tanker is

$$v_2 = 2\cdot78 - 0\cdot05 \times 40$$

$$= 0\cdot78\text{m/s}$$

Time to change positions $= 40s$

Final speed of tug $= 6.78 m/s$

Final speed of tanker $= 0.78 m/s$

Q.23 A bend in a canal follows a circular path of 16km radius. Two barges travelling towards each other have velocities of 11km/hour and 15km/hour respectively. If the distance between the barges, measured in a straight line, is 20km determine the time taken until they are 2km apart.

A.

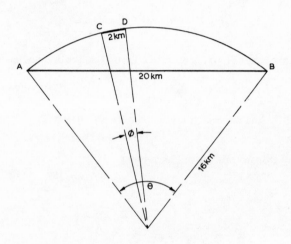

Fig. 23a   Curve of canal and position of barges

Let the initial positions of the barges be A and B and their final positions be C and D.

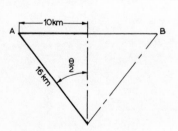

Fig. 23b   Initial range of barges

To find the lengths of arcs AC and BC.

Consider half the angle θ subtended by the chord AB, Fig. 23b

$$\sin\frac{\theta}{2} = \frac{10}{16}$$

$$= 0.625$$

$$\therefore \quad \frac{\theta}{2} = 38.7°$$

and

$$\theta = 77.4°$$

42

Also consider half the angle ϕ subtended by the chord CD, Fig. 23c

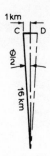

$$\sin\frac{\phi}{2} = \frac{1}{16}$$

$$= 0{\cdot}0625$$

$$\therefore \quad \frac{\phi}{2} = 3{\cdot}58°$$

$$\phi = 7{\cdot}16°$$

Fig. 23c   Final range of barges

The length of the arc AB $= \theta r$ where θ is in radians

$$= 77{\cdot}4 \times \frac{\pi}{180} \times 16$$

$$= 21{\cdot}6\text{km}$$

and the length of arc CD $= \phi r$ where ϕ is in radians

$$= 7{\cdot}16 \times \frac{\pi}{180} \times 16$$

$$= 2\text{km}$$

The distance moved by one barge relative to the other

$$s = 21{\cdot}6 - 2$$

$$= 19{\cdot}6\text{km}$$

and their relative velocity

$$v = 15 + 11$$

$$= 26\text{km/h}$$

The time taken for the movement of each barge

$$t = \frac{s}{v}$$

$$= \frac{19{\cdot}6}{26} \times 60$$

$$= 45{\cdot}3 \text{ minutes}$$

Time for barge movement $= 45{\cdot}3$ minutes

STRENGTH OF MATERIALS

Q.24 The following results were taken from a tensile test on a steel wire 0·5mm diameter and 1·25m long.

Load N	23	30	38	46	54	63	72	81	90
Extension mm	0·33	0·61	0·84	0·99	1·19	1·39	1·52	2·01	2·23

Draw the graph and from it determine the modulus of elasticity. Calculate the extension of a bar of similar material 3·48m long, 10mm diameter supporting a load of 45kN.

A.

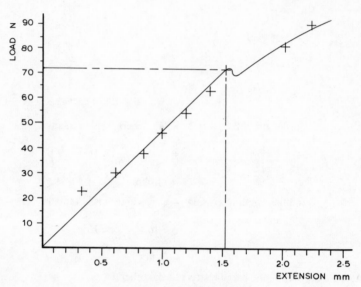

Fig. 24   Load-extension graph

The modulus of elasticity $E = \dfrac{\text{Stress}}{\text{Strain}}$

$$= \frac{\text{Load}}{\text{Area}} \times \frac{\text{Original length}}{\text{Extension}} \qquad \text{(i)}$$

Substituting corresponding values of load and extension from the graph Fig. 24

$$E = \frac{72 \times 4}{\pi \times 0\cdot5^2 \times 10^{-6}} \times \frac{1\cdot25}{1\cdot52 \times 10^{-3}}$$

$$= 301\cdot5 \times 10^9 \text{N/m}^2$$

Transposing equation (i) gives

$$\text{Extension} = \frac{\text{Load}}{\text{Area}} \times \frac{\text{Original length}}{E}$$

$$= \frac{45 \times 10^3 \times 4}{\pi \times 10^2 \times 10^{-6}} \times \frac{3\cdot48}{301\cdot5 \times 10^9}$$

$$= 0\cdot00661\text{m}$$

Modulus of elasticity of wire $= 301\cdot5\text{GN/m}^2$

Extension of bar $\qquad = 6\cdot61\text{mm}$

Q.25 A metal bar of 40mm diameter is sheathed by a sleeve of the same material giving a tight fit. The unit is then subjected to a compressive load. What must be the outside diameter of the sleeve for this to carry 50% of the load carried by the bar?

A.

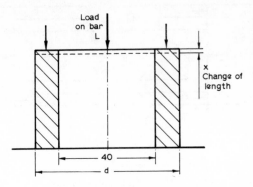

Fig. 25   Bar and sleeve

The sleeve is a tight fit on the bar, therefore, when loaded

The change of length of the bar = Change of length of the sleeve

$$x_B = x_S$$

Now $x = \varepsilon l$

$$\therefore \quad \varepsilon_B = \varepsilon_S \qquad \text{since } l \text{ is common and cancels}$$

and $\varepsilon = \sigma/E$

$$\therefore \quad \sigma_B = \sigma_S \qquad \text{since } E \text{ also is common}$$

Finally $\sigma = F/A$

$$\therefore \quad \frac{F_B}{A_B} = \frac{F_S}{A_S} \qquad \begin{array}{l}\text{where } F_B = L \\ \text{and } F_S = 0.5L\end{array}$$

$$\frac{L \times 4}{\pi \times 40^2} = \frac{0.5L \times 4}{\pi(d^2 - 40^2)}$$

$$d^2 - 40^2 = 0.5 \times 40^2$$

$$d^2 = 800 + 1600$$

$$d = \sqrt{2400}$$

$$= 49\text{mm}$$

Outside diameter of sleeve $\quad = 49\text{mm}$

Q.26 A steel bar 20mm diameter is screwed at each end and fitted with nuts having a pitch of 1mm. A tube of similar material, 1·2m long, having an inside diameter of 30mm and a wall thickness of 5mm is fitted between the nuts and given an initial stress of 12·5kN/m². Determine the initial stress on the bar and the increase in stress in the bar and the tube when one nut is tightened one revolution relative to the other. $E = 210 \times 10^9 \text{N/m}^2$.

A.

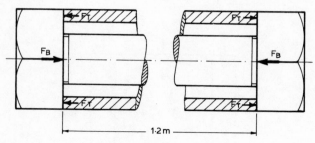

Fig. 26 Bar and tube

The initial load on the bar $F_B = F_T$ the load on the tube.
Now since

$$\sigma = \frac{F}{A}$$

$$F = \sigma A$$

Therefore by substitution

$$\sigma_B A_B = \sigma_T A_T$$

Hence

$$\sigma_B = \sigma_T \times \frac{A_T}{A_B}$$

$$= \frac{12.5 \times 10^3 \times (0.7854)(0.04^2 - 0.03^2)}{(0.7854)(0.02^2)}$$

Using the difference of two squares

$$\sigma_B = \frac{12.5 \times 10^3 \times 0.07 \times 0.01}{0.0004}$$

$$= 12.5 \times 10^3 \times 1.75$$

$$= 21.9 \times 10^3 \text{N/m}^2$$

Since the materials in bar and tube are similar

$$E_B = E_T$$

and since

$$E = \frac{F}{A} \times \frac{l}{x} \quad \text{from } E = \frac{\sigma}{\varepsilon}$$

$$\frac{F_B l_B}{A_B x_B} = \frac{F_T l_T}{A_T x_T}$$

46

Now

$$F_B = F_T$$

and

$$l_B = l_T$$

Therefore

$$A_B x_B = A_T x_T$$

and

$$\frac{x_B}{x_T} = \frac{A_T}{A_B}$$

$$= \frac{(0.7854)(0.04^2 - 0.03^2)}{(0.7854)(0.02^2)}$$

$$= 1.75$$

$$x_B = 1.75 x_T$$

The nut on the bar is revolved once, therefore the total change of length is one pitch

$$x_B + x_T = 0.001$$

$$1.75 x_T + x_T = 0.001$$

$$x_T = \frac{0.001}{2.75}$$

$$= 0.000364$$

and

$$x_B = 0.001 - 0.000364$$

$$= 0.000636$$

From

$$E = \frac{\sigma}{\varepsilon}$$

The increase of stress on tube $\quad \sigma_T = E_T \varepsilon_T$

$$= 210 \times 10^9 \times \frac{0.000364}{1.2}$$

$$= 63.6 \times 10^6 \text{N/m}^2$$

and the increase of stress on bar $\sigma_B = E_B \varepsilon_B$

$$= 210 \times 10^9 \times \frac{0.000636}{1.2}$$

$$= 111.4 \times 10^6 \text{N/m}^2$$

Initial stress on bar	$= 21.9 \text{kN/m}^2$
Increase of stress on tube	$= 63.6 \text{MN/m}^2$
Increase of stress on bar	$= 111.4 \text{MN/m}^2$

47

Q.27 A bottom end bolt 560mm long has an effective diameter of 60mm, a pitch of 4.2mm and a modulus of elasticity of $2 \cdot 1 \times 10^{11} \text{N/m}^2$. The nut is tightened and the bolt is subject to a tensile load of $5 \cdot 1 \times 10^6 \text{N}$. If 50% of the resulting deformation is carried by the bolt find the angular movement of the nut in degrees.

A. Using the standard equations

$$E = \frac{\sigma}{\varepsilon}$$

$$\sigma = \frac{F}{A}$$

and

$$\varepsilon = \frac{x}{l}$$

The change of length of the bolt

$$x = \varepsilon l$$

$$= \frac{\sigma l}{E}$$

$$= \frac{Fl}{AE}$$

$$= \frac{5 \cdot 1 \times 10^6 \times 0 \cdot 56}{(0 \cdot 7854) \times 0 \cdot 06^2 \times 2 \cdot 1 \times 10^{11}}$$

$$= 4 \cdot 82 \times 10^{-3} \text{m}$$

This represents 50% of the total deformation

$$\therefore \quad \text{Total deformation} = 4 \cdot 82 \times 10^{-3} \times 2$$

$$= 9 \cdot 64 \times 10^{-3} \text{m}$$

This deformation in the bolt and the bearing is caused by rotating the nut; one revolution causing a deformation of 4.2mm. To obtain the deformation required the nut must revolve

$$\frac{9 \cdot 64 \times 10^{-3}}{4 \cdot 2 \times 10^{-3}} = 2 \cdot 29 \text{rev}$$

$$= 2 \cdot 29 \times 360$$

$$= 825 \text{ degrees}$$

Q.28 A horizontal beam is suspended from two similar metal bars, 12mm diameter. One bar is 6m long and the other is 4.5m long. A load of 10kN is placed on the beam in such a position that the beam remains horizontal. Find the stresses set up in the bar.

A.

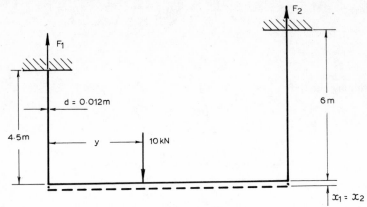

Fig. 28   Horizontal beam

For the beam to remain horizontal after loading the change of length of each bar must be the same, hence

$$x_1 = x_2$$

Now

$$\varepsilon = \frac{x}{l}$$

thus

$$\varepsilon_1 l_1 = \varepsilon_2 l_2$$

also

$$E = \frac{\sigma}{\varepsilon}$$

thus

$$\frac{\sigma_1 l_1}{E_1} = \frac{\sigma_2 l_2}{E_2}$$

Finally

$$\sigma = \frac{F}{A}$$

Therefore

$$\frac{F_1 l_1}{E_1 A_1} = \frac{F_2 l_2}{E_2 A_2}$$

The bars are of similar material and diameter, hence E and A are common factors, thus

$$F_1 l_1 = F_2 l_2 \qquad\qquad \text{(i)}$$

49

Equating the forces for vertical equilibrium

$$F_1 + F_2 = 10\text{kN}$$

thus

$$F_2 = 10 - F_1$$

Substituting into (i)

$$F_1 l_1 = (10 - F_1)l_2$$

$$F_1 l_1 + F_1 l_2 = 10 l_2$$

$$\text{Force on bar 1, } F_1 = \frac{10 l_2}{l_1 + l_2}$$

$$= \frac{10 \times 6}{4 \cdot 5 + 6}$$

$$= 5 \cdot 71\text{kN}$$

$$\text{Force on bar 2, } F_2 = 10 - 5 \cdot 71$$

$$= 4 \cdot 29\text{kN}$$

$$\text{Stress on bar 1, } \sigma_1 = \frac{5 \cdot 71 \times 10^3 \times 4}{\pi \times 0 \cdot 012^2}$$

$$= 50 \cdot 5 \times 10^6 \text{N/m}^2$$

$$\text{Stress on bar 2, } \sigma_2 = \frac{4 \cdot 29 \times 10^3 \times 4}{\pi \times 0 \cdot 012^2}$$

$$= 37 \cdot 9 \times 10^6 \text{N/m}^2$$

$$\text{Stress on bar 1} \quad = 50 \cdot 5\text{MN/m}^2$$

$$\text{Stress on bar 2} \quad = 37 \cdot 9\text{MN/m}^2$$

Q.29 Two shafts, one brass the other steel, are subject to axial load. If the strain on the brass is 25% greater than the strain on the steel when twice the load is applied, determine the ratio of their diameters.

$$\text{Modulus of Elasticity for Brass} = 88\text{GN/m}^2$$
$$\text{Modulus of Elasticity for Steel} = 210\text{GN/m}^2$$

A.

Let the strain on the steel bar $\quad= \varepsilon$
then the strain on the brass bar $\quad= 1 \cdot 25\varepsilon$
also let the load on the steel bar $\quad= F$
then the load on the brass bar $\quad= 2F$
Finally let the area of the steel bar $= 0 \cdot 7854 D_S^2$
and the area of the brass bar $\quad= 0 \cdot 7854 D_B^2$

Using the equation for stress

$$\sigma = \frac{F}{A}$$

and substituting into

$$E = \frac{\sigma}{\varepsilon}$$

gives

$$E = \frac{F}{A\varepsilon}$$

Transposing this equation gives

$$A = \frac{F}{E\varepsilon} \qquad \text{(i)}$$

Substituting the above figures into equation (i) gives for the steel bar

$$0.7854D_S^2 = \frac{F}{210 \times 10^9 \times \varepsilon}$$

hence

$$D_S^2 = \frac{F}{0.7854 \times 210 \times 10^9 \times \varepsilon}$$

and for the brass bar

$$0.7854D_B^2 = \frac{2F}{88 \times 10^9 \times 1.25\varepsilon}$$

hence

$$D_B^2 = \frac{2F}{0.7854 \times 88 \times 10^9 \times 1.25\varepsilon}$$

For the ratio of diameters $\dfrac{D_S^2}{D_B^2} = \dfrac{F}{0.7854 \times 210 \times 10^9 \times \varepsilon} \times \dfrac{0.7854 \times 88 \times 10^9 \times 1.25\varepsilon}{2F}$

$$\left(\frac{D_S}{D_B}\right)^2 = \frac{88 \times 1.25}{210 \times 2}$$

$$\frac{D_S}{D_B} = \sqrt{0.262}$$

$$= 0.512$$

The ratio of diameters $\dfrac{D_S}{D_B} = 0.512$.

Q.30 A uniform beam 4·4m long is supported by two props of equal length and a hinge at its left-hand end. One prop positioned 1·4m from the left-hand end has a cross-sectional area of 0·03m^2 and a modulus of elasticity of 210×10^6kN/m^2. The other prop is positioned 3·5m from the left-hand end, has a cross-sectional area of 0·05m^2

and a modulus of elasticity of $175 \times 10^6 kN/m^2$. The beam is loaded centrally with a mass of 100kg and remains horizontal while the load is applied. Determine the load on each prop and the hinge.

A.

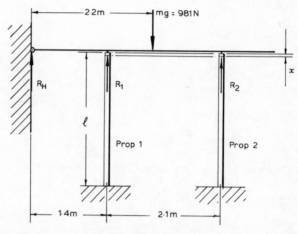

Fig. 30   Horizontal beam

The beam remains horizontal during loading, thus the change of length of each prop must be equal. Since the length of each prop is also equal it follows that the strain

$$\frac{x}{l} = \text{a constant } \varepsilon$$

Using the equations

$$E = \frac{\sigma}{\varepsilon}$$

and

$$\sigma = \frac{F}{A}$$

gives

$$F = EA\varepsilon \qquad \text{(i)}$$

Substituting known values into equation (i) gives the reaction in prop 1

$$R_1 = 210 \times 10^9 \times 0\cdot03 \times \varepsilon$$
$$= 6\cdot3 \times 10^9 \varepsilon \qquad \text{(ii)}$$

and the reaction in prop 2

$$R_2 = 175 \times 10^9 \times 0\cdot05 \times \varepsilon$$
$$= 8\cdot75 \times 10^9 \varepsilon \qquad \text{(iii)}$$

Taking moments about the hinge and equating for equilibrium

$$2.2 \times mg = 1.4 \times R_1 + 3.5 \times R_2$$
$$= (1.4 \times 6.3 \times 10^9 + 3.5 \times 8.75 \times 10^9)\varepsilon$$

hence

$$\varepsilon = \frac{2.2 \times 100 \times 9.81}{(1.4 \times 6.3 + 3.5 \times 8.75)10^9}$$
$$= 54.8 \times 10^{-9}$$

Substituting into equation (ii)

$$R_1 = 6.3 \times 10^9 \times 54.8 \times 10^{-9}$$
$$= 345N$$

and into equation (iii)

$$R_2 = 8.75 \times 10^9 \times 54.8 \times 10^{-9}$$
$$= 480N$$

The reaction at the hinge $R_H = mg - 345 - 480$
$$= 981 - 825$$
$$= 156N$$

The load on prop 1	$= 345N$
The load on prop 2	$= 480N$
and the load on the hinge	$= 156N$

Q.31 A straight steel steam pipe 5m long is fitted between two rigid bulkheads. When heated to its working temperature its total free expansion would be 13·5mm. If the compressive stress in the pipe is to be limited to $4.5MN/m^2$, calculate the tensile stress to be exerted on the pipe when fitted cold. Take the modulus of elasticity of steel as $207GN/m^2$.

A. At working temperature, without pre-stressing and using

$$E = \frac{\text{Stress}}{\text{Strain}}$$

The stress induced

$$\sigma = E \times \varepsilon$$

and since the strain

$$\varepsilon = \frac{x}{l}$$

then the induced stress $\qquad \sigma = \frac{Ex}{l}$

$$= 207 \times 10^9 \times \frac{0.0135}{5}$$
$$= 559 \times 10^6 N/m^2$$

Since the limiting stress on the pipe is $415 \times 10^6 \text{N/m}^2$

$$\text{then the initial tensile stress} = (559 - 415)10^6$$

$$= 144 \times 10^6 \text{N/m}^2$$

$$\text{Initial tensile stress required} = 144 \times 10^6 \text{N/m}^2$$

Q.32 A brass bar 200mm diameter is cooled from 20°C to minus 10°C. The modulus of elasticity for brass is $88 \times 10^9 \text{N/m}^2$, and the coefficient of linear expansion per degree is 5×10^{-6}. Determine:
  (a) The extension per unit length due to free contraction.
  (b) The force in the bar if contraction is prevented.

A. The change in length of the bar due to change of temperature is given by

$$x = \alpha \times l \times \delta\theta \qquad \text{where } \delta\theta = t_2 - t_1$$

$$= 5 \times 10^{-6} \times 1 \times (-10 - 20)$$

$$= -150 \times 10^{-6} \text{ per unit length}$$

Now since

$$E = \frac{\sigma}{\varepsilon}$$

$$\sigma = \frac{F}{A}$$

and

$$\varepsilon = \frac{x}{l}$$

then

$$E = \frac{F \times l}{A \times x}$$

and the force in the bar $\qquad F = \dfrac{E \times A \times x}{l}$

$$= \frac{88 \times 10^9 \times \pi \times 0.2^2 \times (-150 \times 10^{-6})}{4 \times 1}$$

$$= 415 \times 10^3 \text{N}$$

The extension of bar $\qquad = -150 \times 10^{-6}$ per unit length

The force to prevent contraction $= 415\text{kN}$

Q.33 A boiler tube of 60mm outside diameter is subject to an internal pressure of 4MN/m^2. The maximum hoop stress in the tube wall is limited to 20MN/m^2. Working from first principles, find the wall thickness.

A.

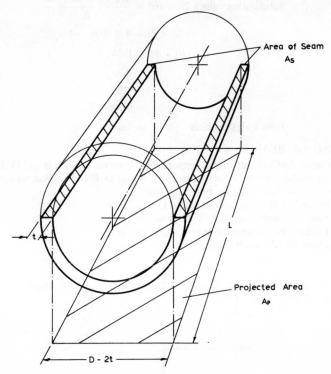

Fig. 33   Boiler tube

Consider two halves of the boiler tube divided by longitudinal seams. The force tending to separate the two halves is given by

$$F = p \times A_P$$

where

$$A_P = (D - 2t)L$$

The area of tube material supporting this load is

$$A_S = 2t \times L$$

and the stress on the longitudinal seam

$$\sigma = \frac{F}{A_S}$$

$$= \frac{p \times A_P}{A_S}$$

$$= \frac{p \times (D - 2t)L}{2t \times L}$$

55

$$\text{Substituting values } 20 \times 10^6 = \frac{4 \times 10^6(0.06 - 2t)}{2t}$$

$$20 \times 2t = 4 \times 0.06 - 4 \times 2t$$

$$(40 + 8)t = 0.24$$

$$t = \frac{0.24}{48}$$

$$= 0.005\text{m}$$

Tube wall thickness $\quad\quad = 5\text{mm}$

BENDING OF BEAMS

Q.34 A beam ABCDE 8m long is supported 2m from each end at B and D. Two loads of 4kN are uniformly distributed between A–B and D–E and a concentrated load of 5kN is placed halfway along the beam at point C.

  (a) Determine the reactions at B and D.
  (b) Find the shear force at A, B, C, D and E.
  (c) Find the bending moment at A, B, C, D and E.
  (d) Draw the SF and BM diagrams.

A.

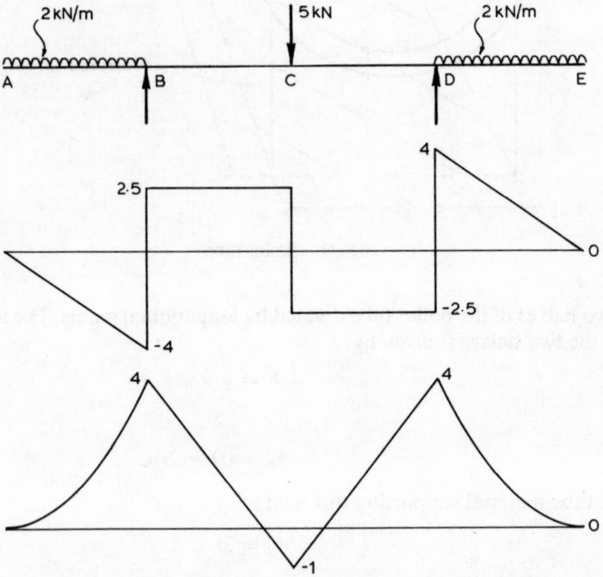

Fig. 34 Symmetrically loaded beam SF and BM diagrams

Since the loading of the beam is symmetrical the reactions at B and D share equally the total load

$$B + D = 4 + 4 + 5$$

$$= 13$$

$$\therefore \quad B = D = 6.5\text{kN}$$

56

The loading of the beam to the left of the following points give the

SF at A	= 0	= 0
SF at B	= −4	= −4kN
SF at C	= −4 + 6·5	= 2·5kN
SF at D	= −4 + 6·5 − 5 + 6·5	= 4kN
SF at E	= −4 + 6·5 − 5 + 6·5 − 4	= 0

Taking moments of forces to the left of these points gives the

BM at A = 0   = 0
BM at B = 4 × 1   = 4kNm
BM at C = 4 × 3 − 6·5 × 2   = −1kNm
BM at D = 4 × 5 − 6·5 × 4 + 5 × 2   = 4kNm
BM at E = 4 × 7 − 6·5 × 6 + 5 × 4 − 6·5 × 2 + 4 × 1 = 0

Q.35 A simply supported beam ABCD is 8m long and supported at B and C. The beam carries a uniformly distributed load of 15kN/m overall and a concentrated load of 40kN at A. If B is positioned 1m from A and both supports carry the same load, at what distance is C from B? Draw shear force and bending moment diagrams to scale and show relevant details.

A.

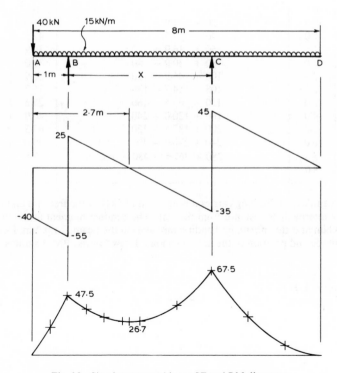

Fig. 35   Simply supported beam SF and BM diagrams

Reactions at B and C share the total load.

$$B + C = (15 \times 8) + 40$$

$$= 160\text{kN}$$

$$\therefore \quad B = C = 80\text{kN}$$

Equating moments of all forces for equilibrium about B

Clockwise moments = anticlockwise moments

$$15 \times 8 \times 3 = 40 \times 1 + 80 \times x$$

Hence

$$80x = 360 - 40$$

$$x = \frac{320}{80}$$

$$= 4\text{m}$$

The table shows the bending moment calculated to the left of any point x measured from the L.H.E. of the beam.

x	$40x + 7{\cdot}5x^2 - 80(x - 1) - 80(x - 5)$	BM
0		0
0·5	20 + 1·9	21·9
1·0	40 + 7·5	47·5
1·5	60 + 16·9 − 40	36·9
2·0	80 + 30·0 − 80	30·0
2·5	100 + 46·9 − 120	26·9
2·7	108 + 54·7 − 136	26·7
3·0	120 + 67·5 − 160	27·5
4·0	160 + 120·0 − 240	40·0
5·0	200 + 187·5 − 320	67·5
6·0	240 + 270·0 − 400 − 80	30·0
7·0	280 + 367·5 − 480 − 160	7·5
8·0		0

Q.36 A cantilever 1·2m long carries a point load of 5kN at the free end and another load some intermediate distance from the wall. The bending moment at the unknown load is 2kNm and the maximum bending moment on the beam is 14kNm. Determine the magnitude and position of the unknown load. Draw SF and BM diagrams.

A.

At x from the L.H.E.

$$\text{BM} = 5 \times x$$

$$= 2\text{kNm}$$

$$\therefore \quad x = \frac{2}{5}$$

$$= 0{\cdot}4\text{m}$$

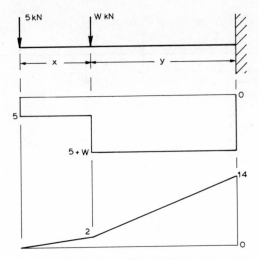

Fig. 36   Cantilever SF and BM diagrams

The cantilever is 1·2m long

$$\therefore \quad y = 1·2 - 0·4$$

$$= 0·8m$$

At 1·2m from L.H.E.

$$\mathbf{BM} = 5 \times 1·2 + W \times 0·8$$

$$= 14kNm$$

$$\therefore \quad W = \frac{14 - 6}{0·8}$$

$$= 10kN$$

Unknown load $= 10kN$

Position from wall $= 0·8m$

Q.37 A beam 8m long is simply supported at its ends and carries a load of 40kN uniformly distributed for five metres from one end. Point loads of 3kN are also carried at each end and at 2m intervals along the beam.
  (a) Determine the reactions at each end.
  (b) Draw to scale the graph of shear forces.
  (c) Find the maximum bending moment.

A.

For equilibrium

Clockwise moments = anticlockwise moments

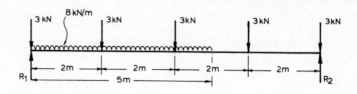

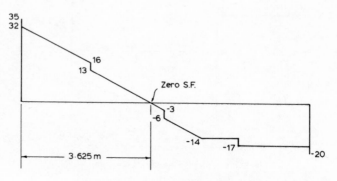

Fig. 37 Simply supported beam SF diagram

Taking moments about R_2

$$8R_1 = 3 \times 8 + 3 \times 6 + 3 \times 4 + 3 \times 2 + 40 \times 5 \cdot 5$$

$$= 24 + 18 + 12 + 6 + 220$$

$$\therefore \quad R_1 = \frac{280}{8}$$

$$= 35 \text{kN}$$

Since the total load $= R_1 + R_2$ then

$$R_1 + R_2 = 3 \times 5 + 8 \times 5$$

$$= 55$$

$$R_2 = 55 - 35$$

$$= 20 \text{kN}$$

The maximum BM occurs at the point of zero SF 3·625m from L.H.E. Taking moments of forces to the left of this point gives

$$\text{Maximum bending moment} = -(32 \times 3 \cdot 625) + (3 \times 1 \cdot 625) + \left(8 \times 3 \cdot 625 \times \frac{3 \cdot 625}{2}\right)$$

$$= -116 + 4 \cdot 875 + 52 \cdot 6$$

$$= -58 \cdot 6 \text{kNm}$$

Reactions at ends of beam $= 35 \text{kN and } 20 \text{kN}$

Maximum bending moment $= 58 \cdot 6 \text{kNm}$

Q.38 A uniform beam is 3m long and simply supported at each end. It carries a uniform distributed load of 730N/m run and a point load of 3420N at 1m from the left-hand end. Determine the value of the upward force required at 1·2m from the left-hand end so that the BM at midspan is zero.

A.

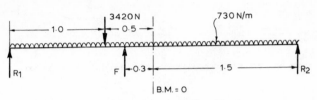

Fig. 38   Uniformly distributed load

Taking moments of forces to the right of the mid-point gives

$$\text{BM at mid-point} = 0$$

$$\therefore \quad \text{Clockwise moments} = \text{anticlockwise moments}$$

$$730 \times 1·5 \times \frac{1·5}{2} = R_2 \times 1·5$$

$$\therefore \quad R_2 = 547 \qquad \qquad \text{(i)}$$

Taking moments of forces to the left of the mid-point also gives

$$\text{BM at mid-point} = 0$$

and again

$$\text{Clockwise moments} = \text{anticlockwise moments}$$

$$R_1 \times 1·5 + F \times 0·3 = 3420 \times 0·5 + 730 \times 1·5 \times \frac{1·5}{2}$$

$$R_1 = \frac{1710 + 822 - 0·3F}{1·5}$$

$$= 1688 - 0·2F \qquad \qquad \text{(ii)}$$

Equating the vertical forces for equilibrium

$$R_1 + F + R_2 = 3420 + 730 \times 3$$

$$\therefore \quad F = 5610 - R_1 - R_2$$

Substituting equations (i) and (ii) gives

$$F = 5610 - 547 - 1688 + 0·2F$$

$$0·8F = 3375$$

$$F = 4420\text{N}$$

$$\text{Upward force required} = 4420\text{N}$$

STRESS IN BEAMS

Q.39 A bar of circular cross-section 200mm diameter and 3m long has a uniformly distributed load of 18kN/m run with a 40kN load at midspan. Calculate the maximum stress on the beam when simply supported at each end.

A.

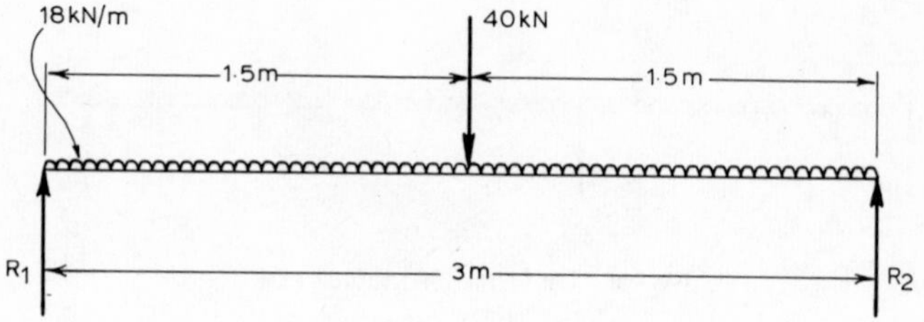

Fig. 39   Beam of circular cross-section

Since the beam is symmetrically loaded

$$R_1 = R_2 = 0.5 \times \text{total load}$$
$$= 0.5(18 \times 3 + 40)$$
$$= 0.5(54 + 40)$$
$$= 47\text{kN}$$

Maximum bending moment occurs at midspan

$$M = 1.5 \times 47 - (1.5 \times 18 \times 0.5 \times 1.5)$$
$$= 70.5 - 20.25$$
$$= 50.25\text{kNm}$$

Maximum bending stress $\sigma = \dfrac{My}{I}$

$$= \frac{50.25 \times 0.1 \times 64}{\pi \times 0.2^4}$$
$$= 64\,000\text{kN/m}^2$$

Maximum stress on beam $= 64\text{MN/m}^2$

Q.40 A piece of timber 0.02m square with simple supports 400mm apart was found to break under a load of 2kN. A beam of similar material 100mm wide and 5m long is required to support 10kN. What minimum depth of material is required?

A.

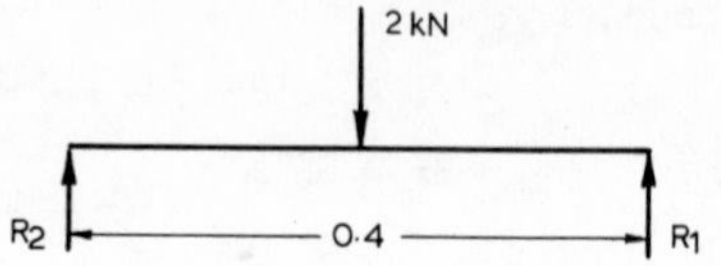

Fig. 40a   Test specimen

The bending moment is maximum when the load is applied at midspan.
The reactions share the total load, hence in Fig. 40a

$$R_1 = R_2$$
$$= 1 \text{kN}$$

The maximum BM $= R_1 \times 0.5l$

$$= 1 \times 0.2$$
$$= 0.2 \text{kNm}$$

From

$$\frac{M}{I} = \frac{\sigma}{y}$$

the limiting stress $\sigma = \dfrac{My}{I}$

$$= 0.2 \times 10^3 \times \frac{0.02}{2} \times \frac{12}{0.02 \times 0.02^3}$$

$$= 150 \times 10^6 \text{N/m}^2$$

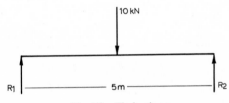

Fig. 40b Timber beam

In Fig. 40b

$$R_1 = R_2$$
$$= 5 \text{kN}$$

The maximum BM $= R_1 \times 0.5l$

$$= 5 \times 0.5 \times 5$$
$$= 12.5 \text{kNm}$$

From

$$\frac{M}{I} = \frac{\sigma}{y}$$

$$\frac{I}{y} = \frac{M}{\sigma}$$

Hence

$$\frac{BD^3}{12} \times \frac{2}{D} = \frac{12.5 \times 10^3}{150 \times 10^6}$$

and

$$D^2 = \frac{12 \cdot 5 \times 10^3}{150 \times 10^6} \times \frac{12}{2 \times 0 \cdot 1}$$

$$D = \sqrt{0 \cdot 005}$$

$$= 0 \cdot 07071 \text{m}$$

Minimum depth of material $= 70 \cdot 71 \text{mm}$

TORSION

Q.41 A hollow shaft of 0·05m O/D and 0·03m I/D transmits power at a speed of 100rev/min. The applied torque of 1540Nm causes the shaft to twist 2° over a length of 3m. Find:
(a) The maximum shear stress on the shaft.
(b) The modulus of rigidity of the material.
(c) The power transmitted.

A.

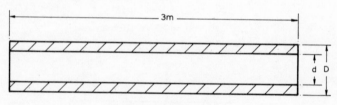

Fig. 41   Hollow shaft

Using the torsion equation

$$\frac{T}{J} = \frac{\tau}{r} = \frac{G\theta}{l}$$

where

$$J = \frac{\pi(D^4 - d^4)}{32}$$

$$r = \frac{D}{2}$$

and

$$\theta = 2° \times \frac{\pi}{180} \text{ radians}$$

The maximum shear stress

$$\tau = \frac{T}{J} \times r$$

$$= \frac{1540 \times 32 \times 0 \cdot 025}{\pi(0 \cdot 05^4 - 0 \cdot 03^4)}$$

$$= 72 \times 10^6 \text{N/m}^2$$

and the modulus of rigidity

$$G = \frac{T}{J} \times \frac{l}{\theta}$$

$$= \frac{1540 \times 32 \times 3 \times 180}{\pi(0.05^4 - 0.03^4) \times 2\pi}$$

$$= 248 \times 10^9 \text{N/m}^2$$

The power transmitted by the shaft $P = T\omega$

where

$$\omega = 2\pi n$$

$$P = 1540 \times 2\pi \times \frac{100}{60}$$

$$= 16.1 \times 10^3 \text{W}$$

Maximum shear stress	$= 72 \text{MN/m}^2$
Modulus of rigidity	$= 248 \text{GN/m}^2$
Power transmitted	$= 16.1 \text{kW}$

Q.42 A torque of 11·3kNm is applied to a solid shaft 125mm diameter and 1·8m long and the angle of twist is one degree. Calculate:
(a) The modulus of rigidity.
(b) The stress in the shaft.
(c) The work done to twist the shaft.

A. Using the torsion equation

$$\frac{T}{J} = \frac{\tau}{r} = \frac{G\theta}{l} \qquad \text{(i)}$$

where

$$J = \frac{\pi D^4}{32}$$

$$r = \frac{D}{2}$$

and

$$\theta = 1° \times \frac{\pi}{180} \text{ radians}$$

From equation (i)
    The modulus of rigidity

$$G = \frac{T}{J} \times \frac{l}{\theta}$$

$$= \frac{11.3 \times 10^3 \times 1.8 \times 32 \times 180}{\pi \times 0.125^4 \times \pi}$$

$$= 48.6 \times 10^9 \text{N/m}^2$$

and the maximum stress $\qquad \tau = \dfrac{T}{J} \times r$

$$= \frac{11 \cdot 3 \times 10^3 \times 32 \times 0 \cdot 125}{\pi \times 0 \cdot 125^4 \times 2}$$

$$= 29 \cdot 5 \times 10^6 \, \text{N/m}^2$$

The work done twisting the shaft $= \dfrac{T\theta}{2}$

$$= \frac{11 \cdot 3 \times 10^3 \times \pi}{2 \times 180}$$

$$= 98 \cdot 6 \, \text{J}$$

The modulus of rigidity $\qquad = 48 \cdot 6 \, \text{GN/m}^2$

The stress in the shaft $\qquad = 29 \cdot 5 \, \text{MN/m}^2$

The work done $\qquad = 98 \cdot 6 \, \text{J}$

Q.43 Two hollow shafts of 400mm outside diameter are connected by a coupling having 6 bolts on a pitch circle diameter of 600mm. The diameter of the bolts is 65mm and the limiting shear stress is the same as the stress on the shaft. Determine the inside diameter of the shafts.

A. The power transmitted by a shaft through a coupling is constant

$\qquad \therefore \quad$ Power transmitted by shaft = Power transmitted by coupling

Since $P = 2\pi n T$ and the rotational speed of shaft and coupling, $2\pi n$, is constant then

$$\text{Torque on shaft} = \text{Torque on coupling}$$

$$T_S = T_C \qquad \qquad \text{(i)}$$

Using the torsion equation

$$\frac{T}{J} = \frac{\tau}{r}$$

the torque on the shaft $\qquad T_S = \dfrac{J_S \tau}{r_S}$

$$= \frac{\pi(0 \cdot 4^4 - d^4) \times \tau}{32 \times 0 \cdot 2} \qquad \qquad \text{(ii)}$$

The torque on the coupling $\quad T_C = F \times R$

where

$$F = \text{force on the bolts}$$

and

$$R = \text{radius of the pitch circle}$$

From the shear stress equation $\tau = \dfrac{\text{Force shearing bolt}}{\text{Cross-sectional area of bolt}}$

the force on one bolt $\qquad = \tau \times A$

thus

$$F = \tau A \times \text{number of bolts}$$

and

$$T_C = \tau A N \times R \qquad\qquad\text{(iii)}$$

Substituting (ii) and (iii) into (i)

$$\frac{\pi(0{\cdot}4^4 - d^4)\tau}{32 \times 0{\cdot}2} = \tau \times A \times N \times R$$

$$0{\cdot}4^4 - d^4 = \frac{\pi \times 0{\cdot}065^2 \times 6 \times 0{\cdot}3 \times 32 \times 0{\cdot}2}{4 \times \pi}$$

$$d^4 = 0{\cdot}0256 - 0{\cdot}0122$$

$$d = \sqrt[4]{0{\cdot}0134}$$

$$= 0{\cdot}34\text{m}$$

Inside diameter of shaft $\qquad = 340\text{mm}$

Q.44 A propeller shaft and coupling transmit 4MW at 2rev/s. The diameter of the shaft is 300mm and the coupling has 8–60mm diameter bolts on a pitch circle diameter of 400mm. Determine the shear stress on:
(a) The shaft.
(b) The coupling bolts.

A.

From the equation of power $\qquad P = 2\pi n T$

the torque on shaft and coupling $T = \dfrac{P}{2\pi n}$

$$= \frac{4 \times 10^6}{2\pi \times 2}$$

$$= 318 \times 10^3\text{Nm}$$

From the torsion equation $\qquad \dfrac{T}{J} = \dfrac{\tau}{r}$

The stress on the shaft $\qquad \tau = \dfrac{Tr}{J}$

$$= \frac{318 \times 10^3 \times 0{\cdot}15 \times 32}{\pi \times 0{\cdot}3^4}$$

$$= 60 \times 10^6\text{N/m}^2$$

The torque transmitted by a coupling

$$T = F \times R$$

where the total force on the bolts $F = \tau \times A \times N$

Thus

$$T = \tau AN \times R$$

and

$$\tau = \frac{T}{ANR}$$

$$= \frac{318 \times 10^3 \times 4}{\pi \times 0.065^2 \times 8 \times 0.2}$$

$$= 60 \times 10^6 \text{N/m}^2$$

Shear stress on shaft $\qquad = 60\text{MN/m}^2$

Shear stress on bolts $\qquad = 60\text{MN/m}^2$

Q.45 A force of 3kN applied at a certain radius is sufficient to twist off a set screw of 25mm diameter. Determine the force applied at twice the original radius that will twist off a screw of 50mm effective diameter if the limiting shear stress is 25 % less than that of the original screw.

A. The shear strength of a screw is measured by the shear stress imposed at failure, thus

$$\tau_2 = \tau_1 \left(\frac{100 - 25}{100} \right)$$

$$\tau_2 = 0.75\tau_1 \tag{i}$$

Using the torsion equation $\qquad \dfrac{T}{J} = \dfrac{\tau}{r}$

$$\tau = \frac{T \times r}{J}$$

thus in equation (i)

$$\frac{T_2 \times r_2}{J_2} = \frac{0.75 T_1 \times r_1}{J_1} \tag{ii}$$

Now

$$T_1 = F_1 \times R_1$$

and

$$T_2 = F_2 \times R_2$$

where

$$R_2 = 2 \times R_1$$

Therefore in equation (ii)

$$\frac{F_2 \times R_2 \times r_2}{J_2} = \frac{0{\cdot}75 \times F_1 \times R_1 \times r_1}{J_1}$$

$$\frac{F_2 \times 2 \times R_1 \times 0{\cdot}025}{0{\cdot}05^4} = \frac{0{\cdot}75 \times 3 \times R_1 \times 0{\cdot}0125}{0{\cdot}025^4}$$

Cancelling all common factors $F_2 = \dfrac{0{\cdot}75 \times 3}{2} \times \dfrac{0{\cdot}0125}{0{\cdot}025} \times \dfrac{0{\cdot}05^4}{0{\cdot}025^4}$

$$= 1{\cdot}125 \times 0{\cdot}5 \times 16$$

$$= 9\text{kN}$$

Force required to shear screw $= 9\text{kN}$

Q.46 Two shafts each transmit the same power. Shaft A is 160mm diameter and revolves at 1rev/s. Shaft B is 120mm diameter and revolves at 4rev/s. If the maximum shear stress imposed on A is 52MN/m², determine the maximum stress on B.

A. Equating the power of each shaft

$$P_A = P_B$$

Using

$$P = 2\pi n T$$

and cancelling common factors

$$n_A \times T_A = n_B \times T_B$$

Now from the torsion equation

$$\frac{T}{J} = \frac{\tau}{r}$$

$$T = \frac{\tau J}{r}$$

thus

$$\frac{n_A \times \tau_A \times J_A}{r_A} = \frac{n_B \times \tau_B \times J_B}{r_B}$$

Hence

$$\tau_B = \frac{n_A}{n_B} \times \frac{J_A}{J_B} \times \frac{r_B}{r_A} \times \tau_A$$

cancelling common coefficients $\tau_B = \dfrac{1}{4} \times \dfrac{0{\cdot}16^4}{0{\cdot}12^4} \times \dfrac{0{\cdot}06}{0{\cdot}08} \times 52 \times 10^6$

$$= 30{\cdot}8 \times 10^6 \text{N/m}^2$$

Maximum stress imposed on B $= 30{\cdot}8\text{MN/m}^2$

Q.47 A shaft of 340mm diameter is turned down to 310mm diameter. The limiting stress for the shaft before and after turning remains at $55 \times 10^6 \text{N/m}^2$. Calculate:
  (a) The power transmitted by the shaft in its original condition when turning at 1·5rev/s.
  (b) The power reduction necessary after the shaft diameter is reduced.

A.   Power transmitted $P = 2\pi nT$

where, from

$$\frac{T}{J} = \frac{\tau}{r}$$

$$T = \frac{\tau J}{r}$$

Thus

$$P = 2\pi n \times \frac{\tau J}{r}$$

$$= 2\pi n\tau \times \frac{\pi D^4}{32} \times \frac{2}{D}$$

$$= \frac{\pi^2 n\tau D^3}{8}$$

Before machining power transmitted

$$P_1 = \frac{\pi^2 \times 1.5 \times 55 \times 10^6 \times 0.34^3}{8}$$

$$= 4 \times 10^6 \text{W}$$

After machining power transmitted

$$P_2 = \frac{\pi^2 \times 1.5 \times 55 \times 10^6 \times 0.31^3}{8}$$

$$= 3.03 \times 10^6 \text{W}$$

Power reduction required

$$= P_1 - P_2$$

$$= 4 \times 10^6 - 3.03 \times 10^6$$

$$= 0.97 \times 10^6 \text{W}$$

Power transmitted by original shaft $= 4\text{MW}$

Power reduction required $= 0.97\text{MW}$

DYNAMICS

Q.48 Two bodies A and B are travelling independently on a straight line in the same direction with A leading. The mass and velocity of A is 3kg and 6m/s respectively. The corresponding values for B being 4kg and 9m/s respectively. Subsequent to collision they separate with a relative velocity of 5m/s. Determine the velocity of each body after collision.

A.

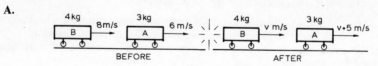

Fig. 48   Trucks before and after collision

70

The theorem of conservation of momentum states

$$\text{Momentum before impact} = \text{Momentum after impact}$$

$$4 \times 8 + 3 \times 6 = 4 \times v + 3(v + 5)$$

$$32 + 18 - 15 = 4v + 3v$$

$$v = \frac{35}{7}$$

$$= 5\text{m/s}$$

$$\text{The velocity of A after impact} = v + 5$$

$$= 10\text{m/s}$$

$$\text{The final velocity of B} \quad = 5\text{m/s}$$

$$\text{and the final velocity of A} \quad = 10\text{m/s}$$

Q.49 A 25 gram projectile travelling at 400m/s passes through a target which is free to move up an inclined track. The projectile leaves the target at 50% of its original velocity and enters a sand bank having a resistance of 10kN. Calculate:
 (a) The vertical distance moved by the 20kg target.
 (b) The depth of penetration into the sand bank.

A.

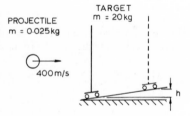
Fig. 49a Projectile and target

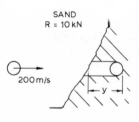
Fig. 49b Projectile and sand

While in contact with the target, the bullet suffers a loss of momentum. From the theorem of conservation of momentum

$$\text{Gain of momentum of target} = \text{loss of momentum of bullet}$$

$$20 \times v = 0{\cdot}025\,(400 - 200)$$

$$\text{Velocity of target after contact } v = 0{\cdot}25\text{m/s}$$

The inclined track converts the kinetic energy available in the moving target into potential energy. Neglecting friction loss at the track

$$KE = PE$$

$$\frac{mv^2}{2} = mgh$$

Therefore

$$h = \frac{v^2}{2g}$$

$$= \frac{0.25^2}{2 \times 9.81}$$

$$= 3.19mm$$

In Fig. 49b the final velocity of the projectile is zero, thus all the remaining kinetic energy is used doing work on the sand

$$KE = \text{Work done}$$

and the work done on the sand = Resistance of sand $\times$ penetration

$$= R \times y$$

$$\therefore \quad Ry = \frac{mv^2}{2}$$

$$y = \frac{0.025 \times 200^2}{2 \times 10 \times 10^3}$$

$$= 0.05m$$

Vertical displacement of target = 3.19mm

Depth of penetration = 50 mm

Q.50 A steel ball of 1.53kg hangs from a fixed point by a cord 2.5m long. If the ball rotates in a horizontal plane at 30 orbits per minute, calculate the diameter of rotation and the tension in the cord.

A.

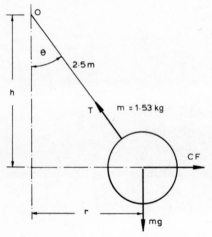

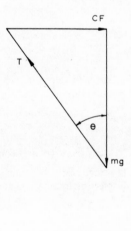

Fig. 50a   Conical pendulum   Fig. 50b   Force vector diagram

To maintain the equilibrium about the point of suspension O in Fig. 50a

$$\text{Clockwise moments} = \text{anticlockwise moments}$$
$$mg \times r = CF \times h$$
$$mgr = m\omega^2 r \times h$$

Since m and r cancel

$$g = \omega^2 h \qquad \text{(i)}$$

where

$$\omega = \frac{2\pi \times 30}{60}$$
$$= \pi \text{rad/s}$$

from (i) the height of orbit $h = \dfrac{g}{\omega^2}$

$$= \frac{9 \cdot 81}{\pi^2}$$
$$= 1\text{m}$$

For the angle of the cord to the vertical θ

$$\cos \theta = \frac{1}{2 \cdot 5}$$
$$= 0 \cdot 4$$
$$\therefore \quad \theta = 66° \ 14'$$

and since

$$\sin \theta = \frac{r}{2 \cdot 5}$$

the radius of orbit $r = 2 \cdot 5 \sin \theta$

$$= 2 \cdot 5 \times 0 \cdot 916$$
$$= 2 \cdot 29\text{m}$$

From the vector diagram Fig. 50b

$$\frac{mg}{T} = \cos \theta$$
$$\therefore \quad T = \frac{mg}{\cos \theta}$$
$$= \frac{1 \cdot 53 \times 9 \cdot 81}{0 \cdot 4}$$
$$= 37 \cdot 5\text{N}$$

Diameter of orbit $\quad = 4 \cdot 58\text{m}$

Tension in the cord $= 37 \cdot 5\text{N}$

Q.51 A 30 gram piece of shrouding breaks away from a turbine rotor at 500mm radius. If the centrifugal forces induced in the rotor are not to exceed 10% of the rotor's weight, determine the maximum speed of the rotor. The mass of the rotor is 10 000kg.

A.

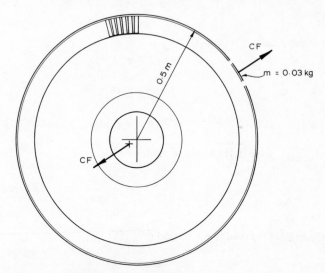

Fig. 51   Turbine rotor

The centrifugal force acting on the piece of shroud in question balances the centrifugal force on the remainder of the rotor thus

$$\text{CF on shroud} = \text{CF on damaged rotor}$$

$$m\omega^2 r = 0.1 \times 10 \times 9810$$

$$\omega^2 = \frac{9810}{0.03 \times 0.5}$$

$$\omega = \sqrt{0.654 \times 10^6}$$

$$= 0.810 \times 10^3 \text{rad/s}$$

but

$$\omega = 2\pi n$$

$$\therefore \quad n = \frac{0.810 \times 10^3}{2\pi}$$

$$= 129 \text{rev/s}$$

Maximum speed of rotor = 129rev/s

Q.52 A mass is placed on a flat disc at a radius of 150mm. Determine the speed of the disc at which the mass is thrown off if the coefficient of friction between the mass and the disc is 0·6. Explain why the mass is subject to an accelerating force when the disc is in uniform motion.

74

A.

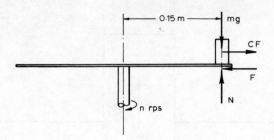

Fig. 52   Rotating disc

The equilibrium of the mass is limited by the speed at which the centrifugal force on the mass is equal to the frictional resistance of the disc.

$$CF = F$$

$$m\omega^2 r = \mu N$$

$$= \mu mg$$

$$\therefore \quad \omega^2 = \frac{\mu g}{r}$$

$$\text{Angular velocity } \omega = \sqrt{\frac{0\cdot6 \times 9\cdot81}{0\cdot15}}$$

$$= \sqrt{39\cdot2}$$

$$= 6\cdot27 \text{rad/s}$$

$$\text{Speed of the disc } n = \frac{\omega}{2\pi}$$

$$= \frac{6\cdot27}{2\pi}$$

$$= 1 \text{rev/s}$$

A body in equilibrium tends to move in a straight line. In this case the rotating disc causes the mass to follow a circular path inducing an accelerating force to change continuously the direction of the mass towards the centre of rotation.

Q.53   A shaft of mass 100kg is supported on two bearings 661mm apart. A flywheel is keyed to the shaft 300mm from the left-hand bearing. When the shaft rotates at 129rev/min the load on the left-hand bearing varies between 1500N and 2700N. Determine:

(a)   The mass of the flywheel.
(b)   The eccentricity of the Centre of Gravity of the flywheel to the axis of rotation.

A.

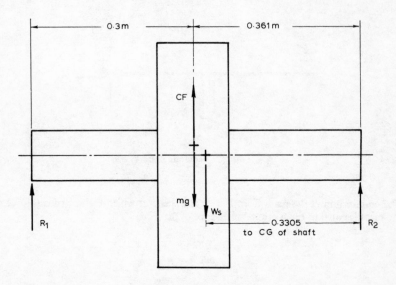

Fig. 53   Shaft and flywheel

The load on the left-hand bearing comprises the static load (say W_1) and the dynamic load (say CF_1).

$$\text{The maximum load } 2700 = W_1 + CF_1$$
$$\text{The minimum load } 1500 = W_1 - CF_1$$
$$\text{Subtracting } 1200 = 2CF_1$$
$$\therefore \quad CF_1 = 600\text{N}$$

i.e. the centrifugal force on the rotating flywheel imposes a dynamic load of 600N on the left-hand bearing

$$\text{The static load on this bearing } W_1 = 2700 - 600$$
$$= 2100\text{N}$$

Equating the moments of static forces for equilibrium about R_2

$$2100 \times 0\cdot661 = mg \times 0\cdot361 + W_S \times 0\cdot3305$$
$$3\cdot54m = 1389 - 324$$
$$\therefore \quad \text{Mass of flywheel } m = \frac{1065}{3\cdot54}$$
$$= 300\text{kg}$$

Equating the moments of dynamic forces about R_2

$$600 \times 0.661 = CF \times 0.361$$

$$\therefore \quad CF = \frac{396.6}{0.361}$$

$$= 1100\text{N}$$

but

$$CF = m\omega^2 r$$

hence

$$r = \frac{CF}{m\omega^2} \qquad \text{(i)}$$

where

$$\omega = \frac{129 \times 2\pi}{60}$$

$$= 13.5\text{rad/s}$$

Substituting the derived values into equation (i)

$$r = \frac{1100}{300 \times 13.5^2}$$

$$= 0.02\text{m}$$

$$\text{Mass of flywheel} = 300\text{kg}$$

$$\text{Eccentricity of C.G.} = 20\text{mm}$$

Q.54 A certain stage in a turbine has brass blades 0.23m long and of regular cross-sectional area. The centre of gravity of the blades is 0.64m from the centre of rotation. Find the stress set up at the root of a blade when the turbine runs at 1500rev/min. The density of brass is 8400kg/m³.

A.

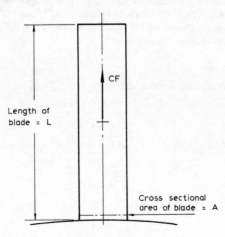

Length of blade = L

CF

Cross sectional area of blade = A

Fig. 54   Turbine blade

The stress on the root of the blade $\sigma = \dfrac{F}{A}$

Where the applied force $F = CF$

$$= m\omega^2 r$$

Now from the equation of density $\rho = \dfrac{m}{V}$

The mass of the blade $m = \rho V$

$$= \rho AL$$

The speed of rotation $\omega = \dfrac{1500 \times 2\pi}{60}$

$$= 157 \text{rad/s}$$

and

the radius of the C.G. $= 0.64$m

Thus by substitution $\sigma = \dfrac{m\omega^2 r}{A}$

$$= \dfrac{\rho AL \times 157^2 \times 0.64}{A}$$

$$= 8400 \times 0.23 \times 24\,650 \times 0.64$$

$$= 305 \times 10^5$$

Stress in turbine blade root $= 30.5 \text{MN/m}^2$

MACHINES

Q.55 A simple machine requires an effort of 57N moving through 3 metres to lift 15kg a distance of 600mm. The same machine is capable of lifting 150kg through 30 metres in 2 minutes. Find the power which must be supplied for the latter operation if the efficiency of the machine remains constant.

A.   The initial load on the machine $= mg$

$$= 15 \times 9.81$$

$$= 147.1 \text{N}$$

The work done on this load $=$ Load $\times$ Distance moved by load

$$= 147.1 \times 0.6$$

$$= 88.3 \text{Nm}$$

The work supplied to machine $=$ Effort $\times$ Distance moved by effort

$$= 57 \times 3$$

$$= 171 \text{Nm}$$

$$\text{Machine efficiency } \eta = \frac{\text{Work output}}{\text{Work input}}$$

$$= \frac{88 \cdot 3}{171}$$

$$= 0 \cdot 516$$

Assuming the efficiency remains constant

$$\text{Work input} = \frac{\text{Work output}}{\eta}$$

$$= \frac{150 \times 9 \cdot 81 \times 30}{0 \cdot 516}$$

$$= 85 \cdot 5 \times 10^3 \text{J}$$

$$\text{Power supplied} = \text{Work input per second}$$

$$= \frac{85 \cdot 5 \times 10^3}{2 \times 60}$$

$$= 712 \cdot 5 \text{W}$$

$$\text{Power required} = 712 \cdot 5 \text{W}$$

Q.56 In an experiment on a square thread screw jack, efforts of 6N, 13N and 19N were required to lift loads of 150N, 375N and 525N respectively. From a graph determine the law of the machine, also determine the velocity ratio if the pitch of the thread is 12mm and the effective length of the toggle bar is 200mm. Find the efficiency of the machine when the load is 300N.

A.

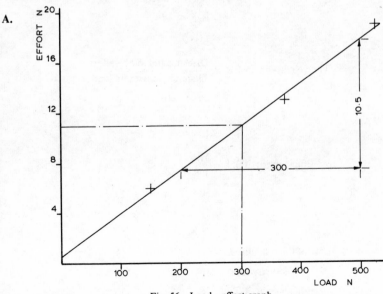

Fig. 56   Load—effort graph

In general the law of a machine is given by

$$E = aW + b$$

where

$$E = \text{the effort}$$
$$W = \text{the load}$$

and

$$a = \text{the slope of the graph}$$
$$b = \text{the intersept}$$

From the graph Fig. 56

$$a = \frac{10 \cdot 5}{300}$$
$$= 0 \cdot 035$$

and

$$b = 0 \cdot 5$$

Thus the law of this machine

$$E = 0 \cdot 035W + 0 \cdot 5$$

With a load of 300N

$$E = 0 \cdot 035 \times 300 + 0 \cdot 5$$
$$= 10 \cdot 5 + 0 \cdot 5$$
$$= 11N$$

$$\text{The velocity ratio } VR = \frac{\text{Distance moved by effort}}{\text{Distance moved by load}}$$
$$= \frac{2\pi \times 200}{12}$$
$$= 104 \cdot 7$$

$$\text{Efficiency at 300N} = \frac{MA}{VR}$$
$$= \frac{300}{11} \times \frac{1}{104 \cdot 7} \times 100$$
$$= 26\%$$

$$\text{Law of the machine } E = 0 \cdot 035W + 0 \cdot 5$$
$$\text{Velocity ratio} = 104 \cdot 7$$
$$\text{Efficiency at 300N} = 26\%$$

Q.57 A screwdown feed check valve acts against a pressure of 15 bars. The valve has a gear ratio of 4:1 and a handwheel of 200mm diameter. The diameter of the valve is 90mm and the spindle has 4 threads per cm. If the efficiency of the entire transmission is 43% find the effort required to close the valve.

A.

$$\text{Pitch of thread} = \tfrac{1}{4} \times 10^{-2}$$

$$= 0{\cdot}0025\text{m}$$

$$\text{The velocity ratio } VR = \frac{\text{Distance moved by effort}}{\text{Distance moved by load}}$$

For one turn of the valve spindle

$$VR = \frac{4 \times \pi \times 0{\cdot}2}{0{\cdot}0025}$$

$$= 1005$$

Combining the three general equations

$$\text{Efficiency } \eta = \frac{MA}{VR}$$

$$\text{Mechanical advantage } MA = \frac{L}{E}$$

and

$$\text{load } L = p \times A$$

gives the equation

$$E = \frac{L}{MA}$$

$$= \frac{p \times A}{\eta \times VR}$$

$$= \frac{15 \times 10^5 \times \pi \times 0{\cdot}045^2}{0{\cdot}43 \times 1005}$$

$$= 22{\cdot}2\text{N}$$

Effort required at handwheel $= 22{\cdot}2\text{N}$

Q.58 A differential axle has drum diameters of 110mm and 80mm. It is driven by a single reduction gear in which the larger gear has 40 teeth and is fixed to the axle. The smaller gear has 8 teeth and is driven by a crank of 0·75m radius. If a load of 250kg is lifted by an effort of 50N determine:
   (a)   The overall velocity ratio.
   (b)   The overall efficiency.

A.

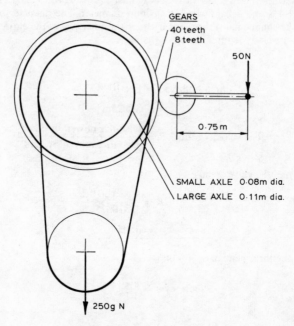

Fig. 58   Gear driven differential axle

Let the axle turn through one revolution. The crank turns through $\frac{40}{8}$ or 5 revolutions and the distance moved by the effort $= 2\pi \times 0.75 \times 5 = 7.5\pi$.

$$\text{The distance moved by the load} = \frac{\pi \times 0.11}{2} - \frac{\pi \times 0.08}{2}$$

$$= 0.015\pi$$

$$\therefore \quad VR = \frac{7.5\pi}{0.015\pi}$$

$$= 500$$

Since

$$\text{mechanical advantage } MA = \frac{L}{E}$$

and

$$\text{efficiency } \eta = \frac{MA}{VR}$$

$$\text{The overall efficiency } \eta = \frac{250 \times 9\cdot81}{50 \times 500} \times 100\%$$

$$= 9\cdot81\%$$

$$\text{Overall velocity ratio} = 500$$

$$\text{Overall efficiency} = 9\cdot81\%$$

Q.59 A lifeboat having a mass of 1000kg is supported on two pulley blocks, the ropes of which lead to a single winch drum. The rope blocks have a single sheave at the lower end and two sheaves at the top. The drum of the winch is 450mm in diameter and the ratio of the driving gears is 30:1. The turning handle has a 300mm crank radius and the overall efficiency of the arrangement is 85%. Find:

    (a) The total velocity ratio.
    (b) The mechanical advantage.
    (c) The effort required to raise the boat.

A.

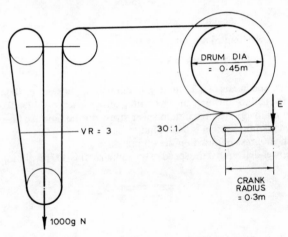

Fig. 59   Lifting blocks and winch

Let the winch drum revolve once then the velocity ratio for the winch

$$= \frac{\text{Distance moved by effort}}{\text{Distance moved by rope}}$$

$$= \frac{30 \times 2\pi \times 0\cdot3}{\pi \times 0\cdot45}$$

$$= 40$$

$$\text{Total velocity ratio } VR = VR \text{ for winch} \times VR \text{ for blocks}$$

$$= 40 \times 3$$

$$= 120$$

From the general equation $\eta = \dfrac{MA}{VR}$

$$MA = \eta \times VR$$
$$= 0.85 \times 120$$
$$= 102$$

and from

$$MA = \dfrac{L}{E}$$

$$E = \dfrac{L}{MA}$$

$$= \dfrac{1000 \times 9.81}{102}$$

$$= 96.2\text{N}$$

Total velocity ratio $= 120$

Mechanical advantage $= 102$

Effort to raise boat $= 96.2\text{N}$

Q.60 A gear train consists of a 72 tooth pinion on an input shaft of 100mm diameter, driving an idler of 20 teeth which engages with another wheel of 24 teeth on the output shaft of 75mm diameter. The maximum shear stress on the input shaft is 13.9MN/m^2 and the efficiency of the system is 0.85. Find:
   (a)   The shear stress imposed on the output shaft.
   (b)   The power delivered if the speed of the input shaft is 300 revolutions per minute.

A.

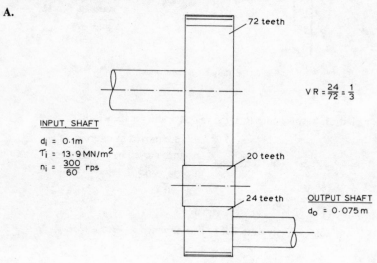

Fig. 60   Gear train

From the torsion equation $\dfrac{T}{J} = \dfrac{\tau}{r}$

the torque on the shaft $T = \dfrac{\tau J}{r}$

Substituting this into the equation for power

$$P = 2\pi n T$$

gives

$$P = 2\pi n \times \frac{\tau J}{r} \tag{i}$$

For the input conditions

$$P_i = 2\pi n_i \times \frac{\tau_i J_i}{r_i}$$

and for the output

$$P_0 = 2\pi n_0 \times \frac{\tau_0 J_0}{r_0}$$

The ratio of these powers $P_i/P_0 = \eta$ the efficiency. From this

$$P_0 = \eta \times P_i \tag{ii}$$

and by substitution

$$2\pi n_0 \frac{\tau_0 J_0}{r_0} = \eta \times 2\pi n_i \frac{\tau_i J_i}{r_i}$$

The stress on the output shaft $\quad \tau_0 = \eta \times \dfrac{n_i}{n_0} \times \tau_i \times \dfrac{J_i r_0}{J_0 r_i} \tag{iii}$

Now the ratio of shaft speeds $n_i/n_0 = VR$ the velocity ratio, where

$$VR = \frac{24}{72}$$

$$= \frac{1}{3}$$

and the ratio

$$\frac{J_i r_0}{J_0 r_i} = \frac{\pi \times d_i^4}{32} \times \frac{d_0}{2} \times \frac{32}{\pi d_0^4} \times \frac{2}{d_i}$$

$$= \frac{d_i^3}{d_0^3}$$

$\therefore$ Substituting into equation (iii)

$$\tau_0 = 0.85 \times \frac{1}{3} \times 13.9 \times 10^6 \times \frac{0.1^3}{0.075^3}$$

$$= 9.335 \times 10^6 \, \text{N/m}^2$$

85

Again combining (i) and (ii)

$$P_0 = \eta \times 2\pi n_i \times \frac{\tau_i J_i}{r_i}$$

$$= 0.85 \times 2\pi \times \frac{300}{60} \times \frac{13.9 \times 10^6}{0.05} \times \frac{\pi \times 0.1^4}{32}$$

$$= 72.9 \times 10^3 \text{W}$$

Shear stress on output shaft $= 9.335 \text{MN/m}^2$

Power transmitted $= 72.9 \text{kW}$

Q.61 A worm driven lifting block consists of a single start worm driving a wheel having 40 teeth. The load drum is 0.25m diameter. If the worm is driven by an electric motor of 3kW turning at 1000rev/min find the maximum load the machine can lift, if its limiting efficiency is 40%.

A.

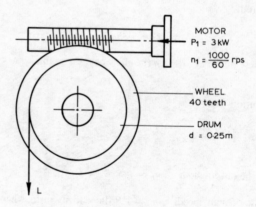

Fig. 61   Worm driven lifting block

If the load drum makes one revolution the single start worm must revolve 40 times. Therefore the speed ratio

$$\frac{n_1}{n_2} = 40$$

and the speed of the drum $n_2 = \dfrac{n_1}{40}$

$$= \frac{1000}{60 \times 40}$$

$$= 0.417 \text{rev/s}$$

The efficiency η $\qquad = \dfrac{P_2}{P_1}$

Therefore power output $\quad P_z = \eta \times P_1$

$$= 0.4 \times 3000$$

$$= 1200\text{W}$$

but

$$P = 2\pi n T$$

and

$$T = L \times r$$

$$\therefore \quad P = 2\pi n L r$$

and the load $\qquad L = \dfrac{P}{2\pi n r}$

$$= \frac{1200}{2\pi \times 0.417 \times 0.125}$$

$$= 3664\text{N}$$

Maximum load on drum $\quad = 3.664\text{kN}$

HYDROSTATICS

Q.62 A cylindrical tank 4m diameter is filled with oil of density 950kg/m^3 to a depth of 3m.

A block of wood of density 750kg/m^3 is floated on the surface and displaces 0.75m^3 of oil. Determine:
   (a) The volume of the block of wood.
   (b) The hydrostatic force on the base of the tank.

A.

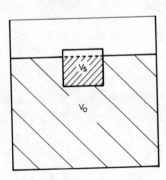

Fig. 62   Wooden block floating in oil

Let the submerged volume of wood $= V_S$ then

$$V_S = 0.75\text{m}^3$$

87

and the volume of the oil $= V_0$

$$= \frac{\pi}{4} \times 4^2 \times 3$$

$$V_0 = 37 \cdot 7 \text{m}^3$$

Using Archimedes Principle for floating bodies

Weight of wood = Weight of oil displaced

and since

$$W = mg \tag{i}$$

and

$$\rho = \frac{m}{V} \tag{ii}$$

then

$$W = V\rho g \tag{iii}$$

and

the weight of wood $= V_s \rho_0 g$

Rearranging equation (iii) gives

$$V = \frac{W}{\rho g}$$

thus

the volume of wood $= \dfrac{V_s \rho_0 g}{\rho_w g}$

$$= \frac{0 \cdot 75 \times 950}{750}$$

$$= 0 \cdot 95 \text{m}^3$$

Hydrostatic force on base of tank = Weight of oil + Weight of wood

$$= V_0 \rho_0 g + V_s \rho_0 g$$

$$= (37 \cdot 7 + 0 \cdot 75)950 \times 9 \cdot 81$$

$$= 359 \text{kN}$$

Volume of wood $= 0 \cdot 95 \text{m}^3$

Force on base of tank $= 359 \text{kN}$

Q.63 A cubic tank of 2m sides is half filled with water of density 1024kg/m³. A wooden cube of 1m sides and density 0·8g/cc is placed in the tank. Find the height of the wood above the water and the rise in the water level.

A.

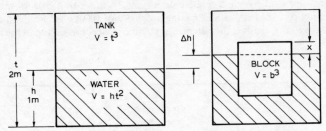

Fig. 63   Changing water level due to displacement by block

Volume of water $ht^2 = 4m^3$

Volume of wood $b^3 = 1m^3$

Mass of wood $m = \rho V$

$$= 0{\cdot}8 \times 1000 \times 1$$

$$= 800kg$$

Where the block is immersed in the water

Weight of water displaced   = Weight of wood

$$V\rho g = mg$$

$$V \times 1024 = 800$$

$\therefore$ Volume of water displaced $V = \dfrac{800}{1024}$

$$= 0{\cdot}781m^3$$

This is equal to the immersed volume of the block

$\therefore$ Depth of block immersed $= \dfrac{V}{b^2}$

$$= \dfrac{0{\cdot}781}{1 \times 1}$$

$$= 0{\cdot}781m$$

Height of wood above water $= 1 - 0{\cdot}781$

$$= 0{\cdot}219m$$

Total volume below water level $= 4 + 0{\cdot}781$

$\therefore$ Final height of water $= \dfrac{4{\cdot}781}{2 \times 2}$

$$= 1{\cdot}195m$$

Change in the water level $= 1{\cdot}195 - 1$

$$= 0{\cdot}195m$$

Q.64 A cube of steel having a relative density of 7·6 floats in a bath of mercury of relative density of 13·6. Fresh water having a density of $1000 kg/m^3$ is added until the steel is just covered. Calculate the percentage volume of steel submerged in the mercury.

A.

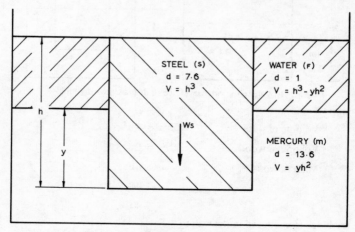

Fig. 64   Steel floating in water and mercury

The cube is in a state of equilibrium, thus its weight is equal to the weight of water and mercury displaced.

$$W_S = W_F + W_M \qquad (i)$$

Using the equations

$$W = mg$$

$$\rho = \frac{m}{V}$$

and

$$d = \frac{\rho}{\rho_F}$$

$$W = dVg\rho_F \qquad (ii)$$

By substituting equation (ii) into equation (i)

$$d_S V_S g \rho_F = d_F V_F g \rho_F + d_M V_M g \rho_F$$

$$7 \cdot 6 h^3 = 1(h^3 - yh^2) + 13 \cdot 6 yh^2$$

$$(7 \cdot 6 - 1)h^3 = (13 \cdot 6 - 1)yh^2$$

$$12 \cdot 6 y = 6 \cdot 6 h$$

$$y = 0 \cdot 524 h$$

Volume of steel submerged in mercury $= 0 \cdot 524$

Percentage volume submerged $\qquad = 52 \cdot 4 \%$

Q.65 A uniform rod with a mass attached at one end floats upright when immersed in a fluid. 200m of the rod shows above the surface when floating in a liquid of S.G. = 0·9. 150mm of the rod is exposed when the liquid S.G. = 0·8. Determine the length of the rod remaining above the surface when floating in a liquid of S.G. = 0·86.

A.

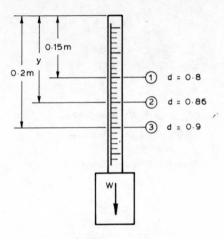

Fig. 65   Floating rod

Let the total volume of body $= V_T$

and the cross-sectional area of rod $= A$

The immersed volume of body $= V_T - A \times$ length of rod above liquid surface

Using the equations

$$W = mg$$

$$\rho = \frac{m}{V}$$

and

$$d = \frac{\rho}{\rho_F}$$

The weight of fluid displaced $W = \rho V g$

$$= d V g \rho_F \qquad \text{(i)}$$

This is constant since the body remains in equilibrium at each relative density condition and its weight remains unaltered.

Comparing conditions 1 and 3

$$W_1 = W_3$$

Substituting (i)

$$d_1 V_1 g \rho_F = d_3 V_3 g \rho_F$$
$$0.8(V_T - 0.15A) = 0.9(V_T - 0.2A)$$
$$(0.9 - 0.8)V_T = (0.18 - 0.12)A$$
$$V_T = \frac{0.06A}{0.1}$$
$$= 0.6A \qquad \text{(ii)}$$

Comparing conditions 1 and 2

$$W_1 = W_2$$

Substituting (i)

$$d_1 V_1 g \rho_F = d_2 V_2 g \rho_F$$
$$0.8(V_T - 0.15A) = 0.86(V_T - yA)$$

Substituting (ii)

$$0.8(0.6A - 0.15A) = 0.86(0.6A - yA) \qquad A \text{ is a common factor}$$
$$0.36 = 0.516 - 0.86y$$
$$y = \frac{0.516 - 0.36}{0.86}$$
$$= 0.1814 \text{m}$$

Length of rod exposed $\qquad = 181.4 \text{mm}$

Q.66 A conical flat-bottomed buoy floats vertically upright in water of density of 1000kg/m^3 with 60% of its axis submerged. The axis is exposed for 1m, the area of the base being 0.6m^2. Determine the mass of lead of density $11\,400 \text{kg/m}^3$ to be externally secured to the base in order to just submerge the buoy.

A.

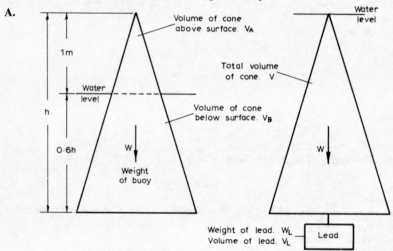

Fig. 66a   Conical buoy   Fig. 66b   Conical buoy with lead attached

For floating bodies Archimedes states:

$$\text{Weight} = \text{Upthrust}$$

and

$$\text{Upthrust} = \text{Weight of fluid displaced}$$

but weight

$$W = mg$$

and mass

$$m = \rho V$$

By combining these equations the weight of a body or of the fluid it displaces may be found

$$W = g\rho V$$

The weight of the buoy $\qquad = \text{weight of fluid displaced (Fig. 66a)}$

$$= g\rho_F V_B$$

The weight of the buoy + lead $= \text{weight of fluid displaced (Fig. 66b)}$

$$= g\rho_F(V + V_L)$$

Hence the weight of the lead $\quad = g\rho_F(V + V_L) - g\rho_F V_B$

$$W_L = g\rho_F(V + V_L - V_B) \qquad \text{(i)}$$

Comparing the similar volumes of V_A and V it is seen that, since

$$h = 1 + 0.6h$$

$$h - 0.6h = 1$$

$$0.4h = 1$$

$$h = 2.5\text{m}$$

the ratio of heights $\qquad = \dfrac{1}{2.5}$

$$= 0.4$$

the ratio of areas $\qquad = 0.4^2$

the ratio of volumes $\qquad = 0.4^3$

Hence

$$V_A = 0.4^3 V$$

Now

$$V = \tfrac{1}{3}h \times \text{Area of base}$$

$$= \frac{2.5 \times 0.6}{3}$$

$$= 0.5\text{m}^3$$

thus

$$V_A = 0.4^3 \times 0.5$$
$$= 0.032 \text{m}^3$$

and

$$V_B = 0.5 - 0.032$$
$$= 0.468 \text{m}^3$$

Using again

$$W = g\rho V$$

the volume of lead $$V_L = \frac{W_L}{g\rho_L}$$

From equation (i)

$$W_L = g\rho_F(V + V_L - V_B)$$
$$= g\rho_F\left(0.5 + \frac{W_L}{g\rho_L} - 0.468\right)$$
$$W_L\left(1 - \frac{g\rho_F}{g\rho_L}\right) = g\rho_F(0.032)$$
$$W_L = \frac{9.81 \times 1000 \times 0.032}{\left(1 - \frac{1000}{11\,400}\right)}$$
$$= \frac{314}{1 - 0.0877}$$
$$= \frac{314}{0.9123}$$
$$= 344 \text{N}$$

The mass of lead
$$= \frac{W_L}{g}$$
$$= \frac{344}{9.81}$$
$$= 35.1 \text{kg}$$

Mass of lead required $\qquad = 35.1 \text{kg}$

Q.67 A fore-peak bulkhead is in the form of an inverted isosceles triangle and separates two fresh water tanks. One tank is filled to a depth of 6m, the other to a depth of 3m. The resultant hydrostatic load on the bulkhead is 7kN. Determine the load on each side.

A. The resultant hydrostatic load is given by $L_1 - L_2$ thus

$$L_1 - L_2 = 7000 \text{ N}$$

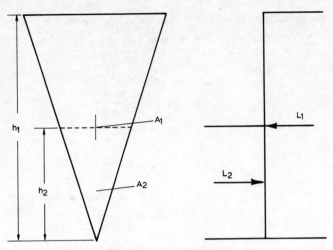

Fig. 67   Fore-peak bulkhead

In general hydrostatic load

$$L = HA\rho g$$

thus

$$H_1 A_1 \rho g - H_2 A_2 \rho g = 7000 \qquad \text{(i)}$$

From Fig. 67 it can be observed that A_1 and A_2 are similar triangles and since

$$\frac{h_1}{h_2} = \frac{6}{3}$$

$$= 2$$

then

$$\frac{A_1}{A_2} = 2^2$$

$$= 4$$

$$\therefore \quad A_1 = 4A_2 \qquad \text{(ii)}$$

The positions of the centroids H are $\frac{1}{3}h$ from the free surface thus

$$H_1 = \frac{h_1}{3}$$

$$= 2\text{m}$$

and

$$H_2 = \frac{h_2}{3}$$

$$= 1\text{m}$$

Substituting equation (ii) and values of H into equation (i) gives

$$(2 \times 4A_2 - 1 \times A_2)\rho g = 7000$$

$$8A_2 - A_2 = \frac{7000}{\rho g}$$

$$7A_2 = \frac{7000}{1000 \times 9\cdot81}$$

$$A_2 = 0\cdot102\text{m}^2$$

Again since

$$L = HA\rho g$$

then

$$L_1 = 2 \times 4 \times 0\cdot102 \times 1000 \times 9\cdot81$$

$$= 8000\text{N}$$

and

$$L_2 = 1 \times 0\cdot102 \times 1000 \times 9\cdot81$$

$$= 1000\text{N}$$

Loads on each side of bulkhead = 8kN and 1kN

Q.68 A dock gate 7·5m wide supports fresh water to a depth of 8m on one side and 3m on the other. Find the resultant force on the gate and the point above the bottom of the gate at which this force acts.

A.

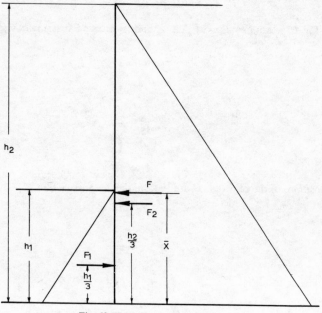

Fig. 68   Loading on dock gate

The hydrostatic load on one side of the dock gate is given by

$$F = HA\rho g$$

This force may be considered to act at the centre of pressure at two-thirds of the total depth of fluid from the free surface.

The resultant hydrostatic load $F = F_2 - F_1$

$$= H_2 A_2 \rho g - H_1 A_1 \rho g$$

$$= [4(8 \times 7 \cdot 5) - 1 \cdot 5(3 \times 7 \cdot 5)]9810$$

$$= [240 - 33 \cdot 75]9810$$

$$= [206 \cdot 25]9810$$

$$= 2023 \text{kN}$$

Equating moments of force about the bottom of the gate

$$F \times \overline{X} = F_2 \times \frac{h_2}{3} - F_1 \times \frac{h_1}{3}$$

$$\therefore \quad \overline{X} = \frac{[240 \times \frac{8}{3} - 33 \cdot 75 \times \frac{3}{3}]9810}{[206 \cdot 25]9810}$$

$$= \frac{640 - 33 \cdot 75}{206 \cdot 25}$$

$$= 2 \cdot 94 \text{m}$$

Resultant force on gate $\qquad = 2023 \text{kN}$

Point at which this force acts $\quad = 2 \cdot 94 \text{m}$ from bottom of gate

Q.69 The side of a tank is formed of a flat triangular plate standing on its base. One side is open to the atmosphere and the other side supports liquid with a free surface at its apex.

Calculate in terms of the vertical height 'h' of the triangle a point where a line parallel to the base can be drawn so that the areas above and below the line are subject to the same hydrostatic force.

A.

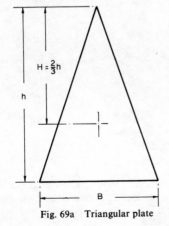

Fig. 69a   Triangular plate

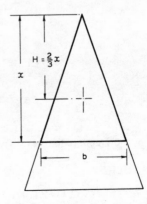

Fig. 69b   Dividing line

Fig. 69a shows the triangular bulkhead and the dimensions required to find the total load acting on its surface

$$\text{Total load on bulkhead} \quad = HA\rho g$$

$$= \frac{2h}{3} \times \frac{hB}{2} \times \rho g \qquad \text{(i)}$$

Fig. 69b shows the horizontal line dividing areas subject to equal load. The load acting above the line is given by

$$\text{Partial load} \quad = HA\rho g$$

$$= \frac{2x}{3} \times \frac{xb}{2} \times \rho g \qquad \text{(ii)}$$

Comparing the base lengths of the similar triangles involved

$$\frac{b}{B} = \frac{x}{h}$$

$$\therefore \quad b = \frac{xB}{h} \qquad \text{(iii)}$$

Since the line divides areas of equal hydrostatic load the partial load above the line equals half the total load.

$$\text{Partial load} = 0.5 \times \text{total load}$$

Substituting (i) and (ii)

$$\frac{2x}{3} \times \frac{xb}{2} \times \rho g = 0.5 \times \frac{2h}{3} \times \frac{hB}{2} \times \rho g$$

Cancelling common factors

$$x^2 b = 0.5h^2 B$$

Substituting equation (iii)

$$x^2 \times \frac{xB}{h} = 0.5h^2 B$$

Hence

$$\frac{x^3}{h^3} = 0.5$$

$$\frac{x}{h} = \sqrt[3]{0.5}$$

$$= 0.794$$

$$\text{Depth of line below free surface } x = 0.794h$$

HYDRAULICS

Q.70 Water flows in the top of a tank at the rate of 60m³/h and out at the bottom through an orifice of 60mm diameter and 0·8 coefficient of discharge. Determine the head of water in the tank for steady conditions.

A.

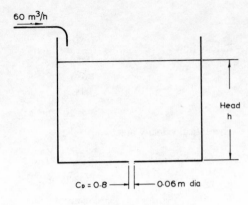

Fig. 70   Water tank

For steady flow conditions the rate of flow through the orifice must equal the rate of flow into the tank.

Rate of flow into tank
$$= 60\text{m}^3/\text{hour}$$
$$= \frac{60}{3600}$$
$$= 0.0167\text{m}^3/\text{s}$$

Rate of flow through orifice
$$= C_D \times \sqrt{2gh} \times 0.785d^2$$
$$= 0.8 \times 0.785 \times 0.06^2 \times \sqrt{2gh}$$
$$= 2.26 \times 10^{-3} \times \sqrt{2gh}$$

Equating for steady flow conditions
$$2.26 \times 10^{-3} \times \sqrt{2gh} = 0.0167$$
$$\sqrt{2gh} = \frac{0.0167}{2.26 \times 10^{-3}}$$

Squaring both sides
$$2gh = 7.37^2$$
$$h = \frac{7.37^2}{2 \times 9.81}$$
$$= 2.76\text{m}$$

Head required for steady conditions = 2.76m

Q.71 A fresh water tank is supplied with 1080kg/hour of water of density 1000kg/m³. Water flows from the tank through an orifice 10mm diameter maintaining a constant head of 3m in the tank. If the coefficient of velocity is 0.97 find:
(a)   The coefficient of contraction of area.
(b)   The coefficient of discharge.

A.

Rate of flow $\qquad V = \dfrac{1080}{1000} \text{m}^3/\text{hour}$

$$= \dfrac{1\cdot080}{3600}$$

$$= 3 \times 10^{-4} \text{m}^3/\text{s}$$

Theoretical velocity of jet $= \sqrt{2gh}$

$$= \sqrt{2 \times 9\cdot81 \times 3}$$

$$= \sqrt{58\cdot86}$$

$$= 7\cdot68 \text{m/s}$$

Actual velocity of jet $\quad v = C_V \times \sqrt{2gh}$

$$= 0\cdot97 \times 7\cdot68$$

$$= 7\cdot45 \text{m/s}$$

Now rate of flow $\qquad V = v \times A$

thus the actual area of jet $= \dfrac{V}{v}$

$$= \dfrac{3 \times 10^{-4}}{7\cdot45}$$

$$= 0\cdot403 \times 10^{-4} \text{m}^2$$

Area of orifice $\qquad = \dfrac{\pi}{4} \times 10^2 \times 10^{-6}$

$$= 0\cdot785 \times 10^{-4}$$

Coefficient of area $\quad C_A = \dfrac{0\cdot403 \times 10^{-4}}{0\cdot785 \times 10^{-4}}$

$$= 0\cdot514$$

Coefficient of discharge $\quad = C_V \times C_A$

$$= 0\cdot97 \times 0\cdot514$$

$$= 0\cdot498$$

Coefficient of area $\qquad = 0\cdot514$

Coefficient of discharge $\quad = 0\cdot498$

Q.72 230kg of fresh water passes through a 20mm orifice in 3 minutes. There is a constant head of 2m above the orifice and the orifice is 2·7m above the ground. The jet of water issuing from the orifice strikes the ground 4·5m from the bottom of the tank. Find C_A, C_V and C_D.

A.

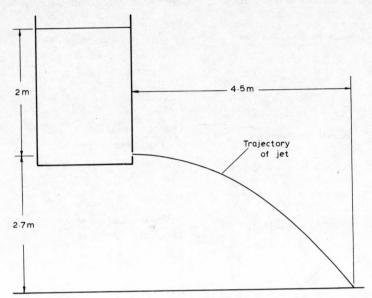

Fig. 72   Jet issuing from tank

Actual rate of discharge $= \dfrac{230}{1000} \times \dfrac{1}{3 \times 60}$

$= 1{\cdot}28 \times 10^{-3}\,\text{m}^3/\text{s}$

Area of the orifice $= 0{\cdot}785d^2$

$= 0{\cdot}785 \times 0{\cdot}02^2$

$= 0{\cdot}314 \times 10^{-3}\,\text{m}^2$

Theoretical velocity of jet $= \sqrt{2gh}$

$= \sqrt{2 \times 9{\cdot}81 \times 2}$

$= 6{\cdot}27\,\text{m/s}$

Theoretical rate of discharge $=$ Area of orifice $\times$ theoretical velocity of jet

$= 0{\cdot}314 \times 10^{-3} \times 6{\cdot}27$

$= 1{\cdot}97 \times 10^{-3}\,\text{m}^3/\text{s}$

Coefficient of discharge $C_D = \dfrac{\text{Actual rate of flow}}{\text{Theoretical rate of flow}}$

$= \dfrac{1{\cdot}28 \times 10^{-3}}{1{\cdot}97 \times 10^{-3}}$

$= 0{\cdot}65$

101

The actual velocity of flow through the orifice is determined from the projected jet.

$$\text{Time for jet to fall } 2 \cdot 7m \qquad t = \sqrt{\frac{2s_V}{g}}$$

$$= \sqrt{\frac{2 \times 2 \cdot 7}{9 \cdot 81}}$$

$$= 0 \cdot 74s$$

$$\text{Actual velocity of jet} \qquad = \frac{\text{Distance travelled}}{\text{Time taken}}$$

$$= \frac{4 \cdot 5}{0 \cdot 74}$$

$$= 6 \cdot 08m/s$$

$$\text{Coefficient of velocity} \qquad C_V = \frac{\text{Actual velocity of jet}}{\text{Theoretical velocity of jet}}$$

$$= \frac{6 \cdot 08}{6 \cdot 27}$$

$$= 0 \cdot 97$$

Now

$$C_D = C_V \times C_A$$

$$\text{Therefore coefficient of area } C_A = \frac{C_D}{C_V}$$

$$= \frac{0 \cdot 65}{0 \cdot 97}$$

$$= 0 \cdot 67$$

$$C_A = 0 \cdot 67$$

$$C_V = 0 \cdot 97$$

$$C_D = 0 \cdot 65$$

Q.73 A rectangular tank having a 300mm diameter orifice at the bottom is filled by means of an on-off cock. The cock is opened when the water level drops to 3·5m and then closed when the level rises to 4m. Coefficient of discharge through the orifice is 0·62 and the free surface area of liquid is 25m². Find the time the cock remains closed between fillings.

A.

$$\text{Rate of discharge through orifice} = C_D \times A \times v$$

Since the velocity of the jet varies with change of head

$$\text{Mean velocity of jet} \qquad v = \frac{v_1 + v_2}{2}$$

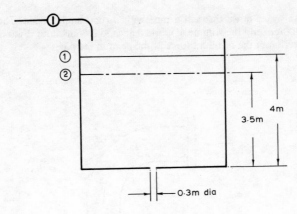

Fig. 73   Water level falls while cock is closed

where
$$v_1 = \sqrt{2gh}$$
$$= \sqrt{2 \times 9\cdot81 \times 4}$$
$$= 8\cdot87\text{m/s}$$

and
$$v_2 = \sqrt{2 \times 9\cdot81 \times 3\cdot5}$$
$$= 8\cdot29\text{m/s}$$

Therefore
$$v = \frac{8\cdot87 + 8\cdot29}{2}$$
$$= 8\cdot58\text{m/s}$$

Rate of discharge $\quad = 0\cdot62 \times 0\cdot785 \times 0\cdot3^2 \times 8\cdot58$
$$= 0\cdot376\text{m}^3\text{/s}$$

While the cock is closed the water level falls from 4m to 3·5m

Volume of water discharged $\quad = 25 \times (4 - 3\cdot5)$
$$= 12\cdot5\text{m}^3$$

Time required for this discharge $\quad = \dfrac{12\cdot5}{0\cdot376}$
$$= 33\cdot2\text{s}$$

Time cock remains closed $\quad = 33\cdot2\text{s}$

103

Q.74 Fresh water flows through a pipe with a right angled bend. The diameter of the pipe is 100mm and the volume of liquid delivered is 36m³/hour. Find the magnitude and the direction of the resultant force at the bend in the pipe.

A.

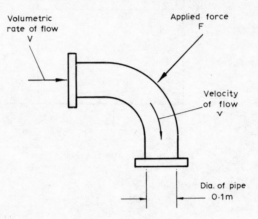

Fig. 74a   Pipe bend

Using the fundamental force equation the direction of flow of water is changed by the applied force F

$$F = ma \tag{i}$$

Considering a time interval of 1 second

$$\text{the mass of water } m \quad = \rho V$$

where

$$\rho = 1000 \text{kg/m}^3$$

and

$$V = \frac{36}{3600}$$

$$= 0 \cdot 01 \text{m}^3/\text{s}$$

The acceleration a is caused by the change of velocity due to the change of direction of flow. The linear speed of flow through the bend is a constant v

$$v = \frac{V}{A}$$

where

$$A = \frac{\pi}{4} \times 0 \cdot 1^2$$

$$= 7 \cdot 85 \times 10^{-3} \text{m}^2$$

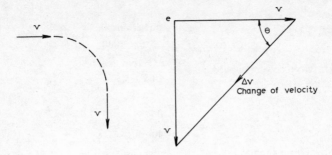

Fig. 74b   Velocity diagram

The acceleration is found using the velocity diagram Fig. 74b.
   In 1 second acceleration

$$a = \Delta v$$
$$= \sqrt{v^2 + v^2}$$
$$= \sqrt{2v^2}$$
$$= 1{\cdot}414v$$

Substituting into equation (i)

$$F = \rho V \times 1{\cdot}414v$$
$$= \rho V \times 1{\cdot}414 \times \frac{V}{A}$$
$$= \frac{1000 \times 0{\cdot}01^2 \times 1{\cdot}414}{7{\cdot}85 \times 10^{-3}}$$
$$= 18\text{N}$$

The direction of the applied force coincides with the acceleration. Referring to Fig. 74b

$$\tan \theta = \frac{v}{v}$$
$$= 1$$

Therefore

$$\theta = 45°$$

Resultant force at bend $= 18\text{N}$ acting at $45°$

Q.75   A 28kw hydraulic motor is supplied with fluid at a pressure of 8MN/m². When used at full power the efficiency of the machine is 70%. Calculate the consumption of fluid.

A.

Efficiency $= \dfrac{\text{power output}}{\text{power input}}$

Therefore power supplied $= \dfrac{\text{Output}}{\text{Efficiency}}$

$= \dfrac{28}{0 \cdot 7}$

$= 40 \text{kW}$

Now power supplied $= \text{pressure} \times \text{volume of fluid/unit time}$

Therefore volumetric flow rate $V = \dfrac{\text{power supplied}}{\text{pressure}}$

$= \dfrac{40 \times 10^3}{8 \times 10^6}$

$= 5 \times 10^{-3} \text{m}^3/\text{s}$

Fluid consumption $= 5 \text{ litres/second}$